Glencoe McGraw-Hill

Math Connects

Course 3

Study Guide and Intervention and Practice Workbook

To the Student This *Study Guide and Intervention and Practice Workbook* gives you additional examples and problems for the concept exercises in each lesson. The exercises are designed to aid your study of mathematics by reinforcing important mathematical skills needed to succeed in the everyday world. The materials are organized by chapter and lesson, with one *Study Guide and Intervention and Practice* worksheet for every lesson in *Glencoe Math Connects, Course 3.*

Always keep your workbook handy. Along with your textbook, daily homework, and class notes, the completed *Study Guide and Intervention and Practice Workbook* can help you review for quizzes and tests.

To the Teacher These worksheets are the same as those found in the Chapter Resource Masters for *Glencoe Math Connects, Course 3.* The answers to these worksheets are available at the end of each Chapter Resource Masters booklet as well as in your Teacher Wraparound Edition interleaf pages.

The McGraw·Hill Companies

Send all inquiries to:
Glencoe/McGraw-Hill
8787 Orion Place
Columbus, OH 43240-4027

ISBN: 978-0-07-881076-3
MHID: 0-07-881076-0

Study Guide and Intervention and Practice Workbook, Course 3

Printed in the United States of America
18 19 20 21 22 23 QVS 20 19 18 17 16

CONTENTS

Lesson/Title **Page**

1-1 Study Guide and Intervention

A Plan for Problem Solving

You can always use the four-step plan to solve a problem.

Understand	Determine what information is given in the problem and what you need to find.
Plan	Select a strategy including a possible estimate.
Solve	Solve the problem by carrying out your plan.
Check	Examine your answer to see if it seems reasonable.

Example 1 **Plant A and Plant B are two new experimental apple trees being grown in a laboratory. The table displays their heights, in millimeters, when they are 5 to 10 days old.**

Day	5	6	7	8	9	10
Plant A	36	39	42	45	48	51
Plant B	32	36	40	44	48	52

Estimate the height of each plant on day 12.

Understand You know their heights for days 5 to 10. You need to determine their heights in two more days.

Plan Determine whether there is a pattern and extend that pattern to day 12.

Solve Comparing each plant's heights on consecutive days, we see that Plant A's height increases by 3 millimeters each day, while Plant B's height increases by 4 millimeters each day. To estimate Plant A's height on day 12, assume that it will grow 3 millimeters each day past day 10, so it will be 51 + 3 + 3 or 57 millimeters. To estimate Plant B's height on day 12, assume that it will grow 4 millimeters each day past day 10, so it will be 52 + 4 + 4 or 60 millimeters.

Check Given what we know about each plant's height and how plants grow in general, both estimates seem reasonable.

Exercises

Use the four-step plan to solve each problem.

1. **MOVIES** A movie ticket costs $3.50. A large popcorn costs $3.75 and a large soda costs $3.00. How much will it cost two friends to go to a movie if they share a popcorn and each has a large soda?

2. **FLOUR BEETLES** The population of a flour beetle doubles in about a week. How long would it take for the population to grow to eight times its original size?

1-1 Practice

A Plan for Problem Solving

Use the four-step plan to solve each problem.

Toppings	Price
1	$12.99
2	$13.79
3	$14.59
4	$15.39

1. **FOOD** The table shows a portion of the price list for a local pizzeria. Tony has $17 that he can spend to buy one large pizza. If the pattern in the prices continues, what is the greatest number of toppings that Tony can order on his pizza? What is the cost of that pizza?

2. **MOVIES** Mr. Sedgwick paid $13 for one adult ticket and one child ticket for a movie. Mrs. Wong paid $18 for one adult ticket and two child tickets to see the same movie, and Mr. Gomez paid $23 for one adult ticket and three child tickets. If the pattern continues, how much should Mrs. Beauregard expect to pay for one adult ticket and four child tickets?

3. **SPORTS** The track coach must buy at least two bottles of water for each participant in a track meet. One team has 35 members, and the other team has 28 members. If each case of water contains 24 bottles, what is the fewest number of full cases that the coach can buy?

4. **GEOGRAPHY** The land area of Washington, D.C., is 61 square miles. In a recent year, the population of Washington, D.C., was 563,384. If one square mile is equal to 640 acres, about how many people per acre were there in Washington, D.C.?

5. **ART SUPPLIES** At the craft store, a paint brush costs $0.79, and a small bottle of paint costs $0.89. What combination of paint brushes and bottles of paint could you buy for exactly $4.15?

6. **GEOMETRY** Draw the next two figures in the pattern.

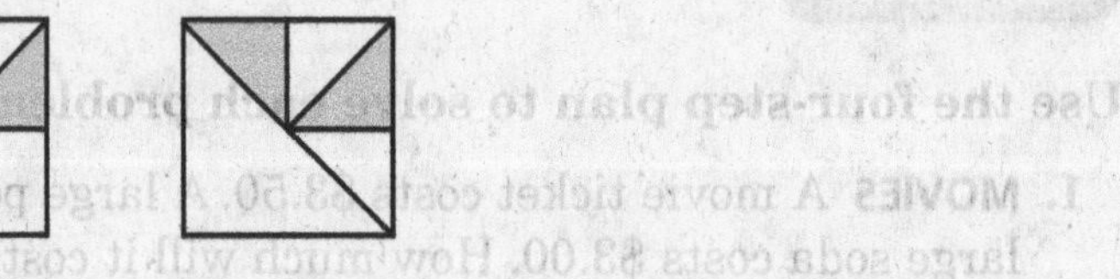

NAME ______________ DATE ______________ PERIOD ______

1-2 Study Guide and Intervention

Variables, Expressions, and Properties

When finding the value of an expression with more than one operation, perform the operations in the order specified by the order of operations.

Order of Operations

1. Perform all operations within grouping symbols first; start with the innermost grouping symbols.
2. Evaluate all powers before other operations.
3. Multiply and divide in order from left to right.
4. Add and subtract in order from left to right.

Example 1 **Evaluate the expression $(5 + 7) \div 2 \times 3 - (8 + 1)$.**

$(5 + 7) \div 2 \times 3 - (8 + 1) = 12 \div 2 \times 3 - (8 + 1)$	Add inside the left parentheses.
$= 12 \div 2 \times 3 - 9$	Add inside the remaining parentheses.
$= 6 \times 3 - 9$	Divide.
$= 18 - 9$	Multiply.
$= 9$	Subtract.

Example 2 **Evaluate the expression $3x^2 - 4y$ if $x = 3$ and $y = 2$.**

$3x^2 - 4y = 3(3)^2 - 4(2)$	Replace *x* with 3 and *y* with 2.
$= 3(9) - 4(2)$	Evaluate the power first.
$= 27 - 8$	Do all multiplications.
$= 19$	Subtract.

Exercises

Evaluate each expression.

1. $4 \times 5 + 8$

2. $16 - 12 \div 4$

3. $14 \div 2 + 3(5)$

4. $5 - 6 \times 2 \div 3$

5. $2 \cdot 3^2 + 10 - 14$

6. $2^2 + 32 \div 8 - 5$

7. $(10 + 5) \div 3$

8. $5^2 \cdot (8 - 6)$

9. $(17 - 5)(6 + 5)$

10. $3 + 7(14 - 8 \div 2)$

11. $5[24 - (6 + 8)]$

12. $\frac{14}{3^2 - 2}$

Evaluate each expression if $a = 3$, $b = 5$, and $c = 6$.

13. $a + 3b$

14. $4b - 3c$

15. $2a - b + 5c$

16. $(ab)^2$

17. $a(b + c)$

18. $3(bc - 8) \div a$

NAME ______________________________ DATE ____________ PERIOD _____

1-2 Practice

Variables, Expressions, and Properties

Evaluate each expression if $r = 3$, $s = 5$, and $t = 2$.

1. $3r + s$

2. $4s - 5t$

3. $8 + 6t - r$

4. rs^2

5. $(st)^2$

6. $\frac{r^2 + 1}{t + 3}$

7. $s(7 + t) - r$

8. $2s^2 - 8s + 3$

Name the property shown by each statement.

9. $6(5 + 1) = 6(5) + 6(1)$

10. $1(2 + 3) = 2 + 3$

11. $(10 + 7) + 4 = 10 + (7 + 4)$

12. $5 + (1 + 9) = 5 + (9 + 1)$

State whether each conjecture is *true* or *false*. If *false*, provide a counter example.

13. The sum of an even number and an odd number is always even.

14. Multiplication of whole numbers is associative.

Rewrite each expression using the indicated property.

15. $(x + 7) + 3$, Associative Property

16. $5(3) + 5(4)$, Distributive Property

17. INTERNET A bookstore offers wireless Internet access to its customers for a charge. The cost of using this service is given by the expression $\$1.50 + \frac{m}{20}$, where m is the number of minutes online. How much would it cost to be online 40 minutes?

18. TEMPERATURE When a temperature in degrees Celsius C is known, the expression $\frac{9C + 160}{5}$ can be used to find the temperature in degrees Fahrenheit. If a thermometer shows that a temperature is 20°C, what is the temperature in degrees Fahrenheit?

NAME ______________________ DATE __________ PERIOD _____

1-3 Study Guide and Intervention

Integers and Absolute Value

A number line can help you order a set of integers. When graphed on a number line, the smaller of two integers is always to the left of the greater integer.

Example 1 **Order the set of integers {10, −3, −9, 4, 0} from least to greatest.**

Graph each integer on a number line.

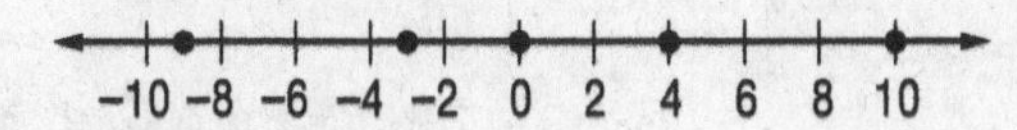

The numbers from left to right are {−9, −3, 0, 4, 10}.

The absolute value of a number is the distance of that number from 0 on a number line.

Example 2 **Evaluate the expression $|-20| + |10|$.**

$|-20| + |10| = 20 + |10|$ The absolute value of −20 is 20.
$= 20 + 10$ The absolute value of 10 is 10.
$= 30$ Simplify.

Exercises

Order each set of integers in each set from least to greatest.

1. {3, 0, −5, 1, 4}

2. {−6, −8, 3, −1, −4}

3. {2, 13, −11, −21, 5}

4. {31, 0, −34, −9, 7}

Evaluate each expression.

5. $|-13|$

6. $|21|$

7. $|-3| + |-5|$

8. $|9| + |-8|$

9. $|-13| + |15|$

10. $|21 - 18|$

11. $|-11| - |-5|$

12. $|4| - |-4|$

13. $|23 + 15|$

Evaluate each expression if $a = -6$, $b = 4$, and $c = 5$.

14. $|a| + 14$

15. $|c - b|$

16. $b + |c|$

17. $|3b|$

18. $2|a| + c$

19. $|2b + c|$

Lesson 1-3

1-3 Practice

Integers and Absolute Value

Replace each ● with $<$, $>$, or $=$ to make a true sentence.

1. 0 ● 8 **2.** −5 ● −3 **3.** 1 ● −7

4. −4 ● −4 **5.** −12 ● 10 **6.** 5 ● −6

7. −6 ● −7 **8.** 0 ● −8 **9.** −10 ● −10

Order each set of integers from least to greatest.

10. {−5, −7, 0, 5, 7} **11.** {−1, 2, −3, 4}

12. {−2, −4, −6, −8, −10, −12} **13.** {0, −9, −3, −7, 1, −1}

Evaluate each expression.

14. $|-19|$ **15.** $|15|$ **16.** $|0|$

17. $|-1| + |3|$ **18.** $|-19| + |-8|$ **19.** $|-12| - |4|$

Evaluate each expression if $k = 4$, $m = -2$, $n = 7$, and $p = -5$.

20. $|m| + 6$ **21.** $n - |p|$ **22.** $k + |p|$

23. $5|n| + k$ **24.** $|n| - 4$ **25.** $9|m| - 14$

TEMPERATURE For Exercises 26 and 28, use the following information.

During a five-day cold spell, Jose recorded the temperature each day at noon. The temperature was −3°F on Monday, −5°F on Tuesday, −4°F on Wednesday, −1°F on Thursday, and 0°F on Friday.

26. On which day was it the coldest at noon?

27. On which day was it the warmest at noon?

28. The temperature at noon on Saturday was 25° warmer than the temperature on Tuesday. What was the temperature on Saturday? Justify your answer using a number line.

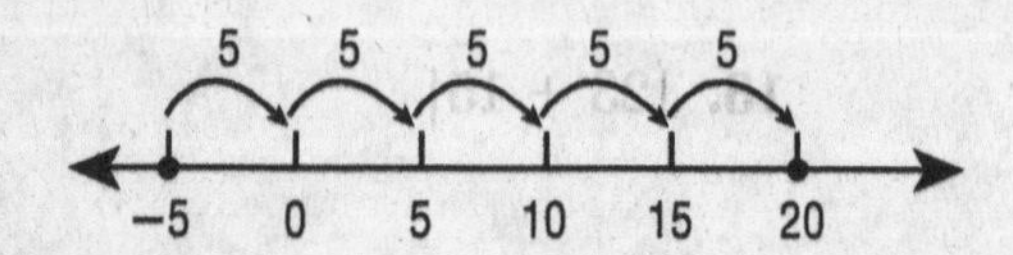

NAME ______________________ DATE ____________ PERIOD ______

1-4 Study Guide and Intervention

Adding Integers

To add integers with the same sign, add their absolute values. The sum has the same sign as the integers.

Example 1 **Find $-3 + (-4)$.**

$-3 + (-4) = -7$ Add $|-3| + |-4|$. Both numbers are negative, so the sum is negative.

To add integers with different signs, subtract their absolute values. The sum has the same sign as the integer with the greater absolute value.

Example 2 **Find $-16 + 12$.**

$-16 + 12 = -4$ Subtract $|12|$ from $|-16|$. The sum is negative because $|-16| > |12|$.

Exercises

Add.

1. $9 + 16$

2. $-10 + (-10)$

3. $18 + (-26)$

4. $-23 + (-15)$

5. $-45 + 35$

6. $39 + (-38)$

7. $-55 + 81$

8. $-61 + (-39)$

9. $-74 + 36$

10. $5 + (-4) + 8$

11. $-3 + 10 + (-6)$

12. $-13 + (-8) + (-12)$

13. $3 + (-10) + (-16) + 11$

14. $-17 + 31 + (-14) + 26$

Evaluate each expression if $x = 4$ and $y = -3$.

15. $11 + y$

16. $x + (-6)$

17. $y + 2$

18. $|x + y|$

19. $|x| + y$

20. $x + |y|$

Lesson 1-4

NAME ______________________________ DATE ____________ PERIOD _____

1-4 Practice

Adding Integers

Find each sum.

1. $-1 + (-8)$ **2.** $13 + 15$ **3.** $19 + (-7)$

4. $-14 + (-14)$ **5.** $-12 + 10$ **6.** $-5 + (-26)$

7. $-46 + 27$ **8.** $-33 + 55$ **9.** $-29 + (-25)$

10. $6 + 14 + (-12)$ **11.** $-15 + (-17) + 10$ **12.** $-13 + (-13) + (-18)$

13. $-5 + 8 + (-1) + (-6)$ **14.** $8 + (-7) + (-8) + (-9)$ **15.** $-15 + 10 + (-16) + 12$

POPULATION For Exercises 16 and 17, use the table below that shows the change in population for four cities between 2000 and 2005.

City	2000 Population (thousands)	Change as of 2005 (thousands)
Boston, Massachusetts	589	−30
Las Vegas, Nevada	478	+67
Pittsburgh, Pennsylvania	335	−18
Rochester, New York	220	−8

Source: U.S. Census Bureau

16. What is the population of each of these cities as of 2005 in thousands?

17. What was the total population change for these four cities?

Write an addition expression to describe each situation. Then find each sum and explain its meaning.

18. GAMES On one turn, you move 10 spaces forward around the game board. On the next turn, you move 4 spaces backward.

19. CAMPING While hiking down into a canyon, Manuel passed a sign stating that the elevation was 100 feet below sea level. He descended another 56 feet before reaching his campsite.

20. WEATHER Before you went to sleep last night, the temperature was −3°F. During the night the temperature dropped by 5°.

21. ELEVATOR Mrs. Brown parked in the parking garage 30 feet below street level. She then got in an elevator and went up 80 feet to her office.

NAME ______________________ DATE __________ PERIOD _____

1-5 Study Guide and Intervention

Subtracting Integers

To subtract an integer, add its opposite or additive inverse.

Example 1 **Find 8 − 15.**

$8 - 15 = 8 + (-15)$ **To subtract 15, add −15.**
$= -7$ **Add.**

Example 2 **Find 13 − (−22).**

$13 - (-22) = 13 + 22$ **To subtract −22, add 22.**
$= 35$ **Add.**

Exercises

Subtract.

1. $-3 - 4$ **2.** $5 - (-2)$ **3.** $-10 - 8$

4. $-15 - (-12)$ **5.** $-23 - (-28)$ **6.** $16 - 9$

7. $9 - 16$ **8.** $-21 - 16$ **9.** $28 - 37$

10. $-34 - (-46)$ **11.** $65 - (-6)$ **12.** $19 - |29|$

Evaluate each expression if $a = -7$, $b = -3$, and $c = 5$.

13. $a - 8$ **14.** $20 - b$ **15.** $a - c$

16. $c - b$ **17.** $b - a - c$ **18.** $c - b - a$

Lesson 1-5

NAME ______________________ DATE ____________ PERIOD _____

1-5 Practice

Subtracting Integers

Subtract.

1. 15 − 7
2. 3 − 12
3. −8 − 9
4. 4 − (−12)
5. 18 − (−7)
6. −8 − (−9)
7. −14 − (−18)
8. −19 − (−13)
9. 8 − (−22)
10. −1 − 15
11. 12 − 19
12. −10 − (−5)

Evaluate each expression if $d = -4$, $f = -7$, and $g = 11$.

13. $d - 10$
14. $g - 15$
15. $d - g$
16. $d - f$
17. $d - f - g$
18. $g - d - f$

GEOGRAPHY For Exercises 19–21, use the table that shows the elevations above sea level of the lowest and highest points on six continents.

Continent	Lowest Point (m)	Highest Point (m)
Africa	−156	5,895
Asia	−400	8,850
Australia	−12	2,228
Europe	−28	5,642
North America	−86	6,194
South America	−42	6,960

19. How far below the highest point in Australia is the lowest point in Australia?
20. How far below the highest point in North America is the lowest point in Asia?
21. Find the difference between the lowest point in South America and the lowest point in Africa.

Simplify.

22. 29 − (−4) − (−15)
23. −10 − [8 + (−16)]
24. 25 − [16 + (−9)]
25. [22 − (−18)] − (−5 + 11)
26. (−5 + 9) − (−20 − 12)
27. [−15 + (−7)] − (−8 − 11)

NAME ______________________________ DATE ____________ PERIOD _____

1-6 Study Guide and Intervention

Multiplying and Dividing Integers

Use the following rules to determine whether the product or quotient of two integers is positive or negative.

- The product of two integers with different signs is negative.
- The product of two integers with the same sign is positive.
- The quotient of two integers with different signs is negative.
- The quotient of two integers with the same sign is positive.

Example 1 **Find 7(−4).**

$7(-4) = -28$ **The factors have different signs. The product is negative.**

Example 2 **Find −5(−6).**

$-5(-6) = 30$ **The factors have the same sign. The product is positive.**

Example 3 **Find 15 ÷ (−3).**

$15 \div (-3) = -5$ **The dividend and divisor have different signs. The quotient is negative.**

Example 4 **Find −54 ÷ (−6).**

$-54 \div (-6) = 9$ **The dividend and divisor have the same sign. The quotient is positive.**

Exercises

Multiply or divide.

1. $8(-8)$ **2.** $-3(-7)$ **3.** $-9(4)$ **4.** $12(8)$

5. $33 \div (-3)$ **6.** $-25 \div 5$ **7.** $48 \div 4$ **8.** $-63 \div (-7)$

9. $(-4)^2$ **10.** $\frac{-75}{15}$ **11.** $-6(3)(-5)$ **12.** $\frac{-143}{-13}$

Evaluate each expression if $a = -1$, $b = 4$, and $c = -7$.

13. $3c + b$ **14.** $a(b + c)$ **15.** $c^2 - 5b$ **16.** $\frac{a-6}{c}$

NAME ______________________ DATE ____________ PERIOD _____

1-6 Practice

Multiplying and Dividing Integers

Multiply.

1. $5(-7)$ **2.** $-3 \cdot 12$ **3.** $-8(-9)$

4. $-4(-12)$ **5.** $(-7)^2$ **6.** $-2(-5)(-3)$

Divide.

7. $-14 \div 2$ **8.** $35 \div (-7)$ **9.** $-48 \div (-6)$

10. $\frac{-66}{6}$ **11.** $\frac{56}{-7}$ **12.** $\frac{-80}{-5}$

Evaluate each expression if $r = -4$, $s = 11$, and $t = -7$.

13. $s + 5t$ **14.** $10 - rt$ **15.** $\frac{5s}{t-4}$

16. $\frac{-42}{r-t}$ **17.** $-r^2 - 16$ **18.** $(2t + 4)^2 \div 4$

Find the mean of each set of integers.

19. $-8, -5, 3, -9, 5, 2$ **20.** $11, -15, -16, 17, -20, -18, -22$

21. $-5, 4, 8, -12, 10$ **22.** $-22, -19, -14, -17, -18$

Find each product or quotient.

23. $(3)^2 \cdot (-4)^2$ **24.** $-3(-5)^2$ **25.** $-5(-2)(4)(-3)$

26. $\frac{-10(15)}{6}$ **27.** $\frac{12^2}{-12}$ **28.** $\frac{-4 \cdot 12}{8}$

29. **MONEY** If you have $216 and you spend $12 each day, how long would it be until you had no money left?

30. **WEATHER** During a six hour period, the temperature dropped 18°F. Find the average hourly change in the temperature.

1-7 Study Guide and Intervention

Writing Equations

The table shows several verbal phrases for each algebraic expression.

Phrases	Expression	Phrases	Expression
8 more than a number the sum of 8 and a number x plus 8 x increased by 8	$x + 8$	the difference of r and 6 6 subtracted from a number 6 less than a number r minus 6	$r - 6$
Phrases	**Expression**	**Phrases**	**Expression**
4 multiplied by n 4 times a number the product of 4 and n	$4n$	a number divided by 3 the quotient of z and 3 the ratio of z and 3	$\frac{z}{3}$

The table shows several verbal sentences that represent the same equation.

Sentences	Equation
9 less than a number is equal to 45. The difference of a number and 9 is 45. A number decreased by 9 is 45. 45 is equal to a number minus 9.	$n - 9 = 45$

Exercises

Write each verbal phrase as an algebraic expression.

1. the sum of 8 and t

2. the quotient of g and 15

3. the product of 5 and b

4. p increased by 10

5. 14 less than f

6. the difference of 32 and x

Write each verbal sentence as an algebraic equation.

7. 5 more than a number is 6.

8. The product of 7 and b is equal to 63.

9. The sum of r and 45 is 79.

10. The quotient of x and 7 is equal to 13.

11. The original price decreased by \$5 is \$34.

12. 5 shirts at \$$d$ each is \$105.65.

1-7 Practice

Writing Equations

Define a variable. Then write an equation to model each situation.

1. After receiving $25 for her birthday, Latisha had $115.

2. At 14 years old, Adam is 3 years younger than his brother Michael.

3. A class of 30 students separated into equal sized teams results in 5 students per team.

4. When the bananas were divided evenly among the 6 monkeys, each monkey received 4 bananas.

Define a variable. Then write an equation that could be used to solve each problem.

5. **GRADES** Kelly's test score was 6 points higher than Michelle's. If Kelly's test score was 88, what was Michelle's test score?

6. **GEOMETRY** A rectangle's width is one-third its length. If the width is 8 inches, what is the length of the rectangle?

7. **FOOTBALL** A team had a total gain of −15 yards over several plays with an average gain of −5 yards per play. How many plays are represented?

Write an equation to model the relationship between the quantities in each table.

8.

Kilograms, k	Grams, g
1	1,000
2	2,000
3	3,000
4	4,000
k	g

9.

Feet, f	Yards, y
3	1
6	2
9	3
12	4
f	y

10. **MONEY** Carlotta earns $3 for every hour that she baby sits. Complete the table of values showing the amount she earns for baby sitting 1, 2, 3, 4, and h hours. Given h, a number of hours, write an equation to find a, the amount that Carlotta earns.

Hours, h	Amount, a

1-8 Study Guide and Intervention

Problem-Solving Investigation: Work Backward

You may need to work backward to solve a problems.

Understand • Determine what information is given in the problem and what you need to find.
Plan • Select a strategy including a possible estimate.
Solve • Solve the problem by carrying out your plan.
Check • Examine your answer to see if it seems reasonable.

Example 1

Mari put money in her savings account each week. She put a certain amount of money in the bank on the first week. On the second week she put twice as much money in the bank as the first week. On the third week, she put \$40 less in the bank than on the second week. On the fourth week, she put \$20 more in the bank than on the third week. Mari put \$200 in the bank on the fourth week. How much money did Mari put in the bank on the first week?

Understand You know that Mari put \$200 in the bank on the fourth week. You need to know how much money she put in the bank on the first week.

Plan Start with the amount she put in the bank on the last week and work backward.

Solve Start with the \$200 Mari put in the bank on the fourth week.

Fourth Week		Third Week		Second Week		First Week
\$200	−\$20	\$180	+\$40	\$220	÷ 2	\$110
This is \$20 more than the third week.	Work backward. Subtract \$20.	This is \$40 less than the second week.	Work backward. Add \$40.	This is twice as much as the first week.	Work backward. Divide by 2.	

Check Start with \$110 for the first week and work forward. On the second week she deposited twice as much money in the bank than on the first week, which is \$220. On the third week, she deposited \$40 less than the second week, which is \$180. On the fourth week she deposited \$20 more than on the third week, or \$200. This is what you know she deposited on the fourth week.

Exercises

Use the work backward strategy to solve each problem.

1. **SHOPPING** Jack spent a total of \$87.58 when he went shopping for camping supplies. He spent \$36.89 on food, \$23.24 on a sleeping bag, and bought lunch. When he got home, he had \$15.70. How much did he spend on lunch?

2. **AGE** Sam is 4 years older than Eliot. Eliot is 9 years younger than Xing. Xing is 3 years older than Damien. If Damien is 15 years old, how old are each of the other boys?

NAME ______________________________ DATE ____________ PERIOD _____

1-8 Practice

Problem-Solving Investigation: Work Backward

Mixed Problem Solving

Use the work backward strategy to solve Exercises 1 and 2.

1. **TRAVEL** Rajiv and his family left home on a trip and drove for 2 hours before they stopped to eat. After 1.5 hours, they were back on the road. They arrived at their destination 3 hours later at 5:00 P.M. What time did they leave home?

2. **GRADES** Kumiko had an average of 92 on her first three math tests. Her scores on the second and third tests were 97 and 89. What was her score on the first test?

Use any strategy to solve Exercises 3–6. Some strategies are shown below.

Problem-Solving Strategies
• Work backward. • Find a pattern.

3. **BAKING** Isabel doubled her recipe for chocolate chip cookies. After her brothers ate 8 cookies, she set aside half of the remaining cookies for a school party. Isabel then gave 2 dozen cookies to her neighbor. She had 12 cookies left over. How many cookies does one recipe make?

4. **ANALYZE TABLES** The table below gives the results from a poll taken at school about the times in minutes that boys and girls spend using the Internet for school work and the total time spent using the Internet each week.

Gender	Time Used for School Work	Total Time per Week
Boys	33 min	255 min
Girls	72 min	213 min

How many more minutes per week do boys spend using the Internet for purposes other than school work than girls?

5. **MOVIES** The two animated films with the highest box office receipts brought in a total of $775 million. If one film brought in $97 million more than the other, how much did the film with the highest receipts bring in?

6. **U.S. PRESIDENTS** Harry S Truman became President on April 12, 1944. He was born on May 8, 1884. How old was he at the time he became President?

NAME ______________________________ DATE ____________ PERIOD ______

1-9 Study Guide and Intervention

Solving Addition and Subtraction Equations

You can use the following properties to solve addition and subtraction equations.

- *Addition Property of Equality*— If you add the same number to each side of an equation, the two sides remain equal.
- *Subtraction Property of Equality*— If you subtract the same number from each side of an equation, the two sides remain equal.

Example 1 **Solve $w + 19 = 45$. Check your solution.**

	$w + 19 = 45$	**Write the equation.**
	$w + 19 - 19 = 45 - 19$	**Subtract 19 from each side.**
	$w = 26$	**$19 - 19 = 0$ and $45 - 19 = 26$. *w* is by itself.**
Check	$w + 19 = 45$	**Write the original equation.**
	$26 + 19 \stackrel{?}{=} 45$	**Replace *w* with 26. Is this sentence true?**
	$45 = 45$ ✓	**$26 + 19 = 45$**

Example 2 **Solve $h - 25 = -76$. Check your solution.**

	$h - 25 = -76$	**Write the equation.**
	$h - 25 + 25 = -76 + 25$	**Add 25 to each side.**
	$h = -51$	**$-25 + 25 = 0$ and $-76 + 25 = -51$. *h* is by itself.**
Check	$h - 25 = -76$	**Write the original equation.**
	$-51 - 25 \stackrel{?}{=} -76$	**Replace *h* with -51. Is this sentence true?**
	$-76 = -76$ ✓	**$-51 - 25 = -51 + (-25)$ or -76**

Exercises

Solve each equation. Check your solution.

1. $s - 4 = 12$ **2.** $d + 2 = 21$ **3.** $h + 6 = 15$

4. $x + 5 = -8$ **5.** $b - 10 = -34$ **6.** $f - 22 = -6$

7. $17 + c = 41$ **8.** $v - 36 = 25$ **9.** $y - 29 = -51$

10. $19 = z - 32$ **11.** $13 + t = -29$ **12.** $55 = 39 + k$

13. $62 + b = 45$ **14.** $x - 39 = -65$ **15.** $-56 = -47 + n$

NAME ______________________ DATE __________ PERIOD _____

1-9 Practice

Solving Addition and Subtraction Equations

Solve each equation. Check your solution.

1. $t + 7 = 12$ **2.** $h - 3 = 8$ **3.** $8 = b - 9$

4. $k - 4 = -14$ **5.** $m + 9 = -7$ **6.** $y - 10 = -3$

7. $-14 = 2 + d$ **8.** $15 + n = 10$ **9.** $-8 = r - 6$

10. $11 = w - 5$ **11.** $-9 = g + 9$ **12.** $12 + c = 16$

13. GEOMETRY Two angles are supplementary if the sum of their measures is 180°. The two angles shown are supplementary. Write and solve an equation to find the measure of angle *R*.

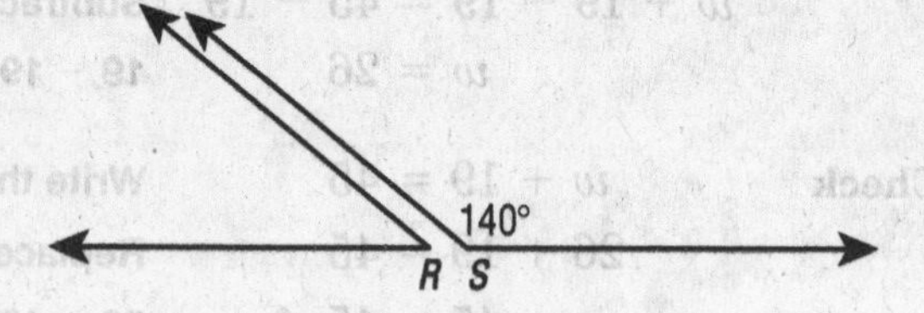

14. ARCHITECTURE The Sears Tower in Chicago was the tallest building in the world when it was completed. Twenty-three years later, a taller building was completed in 1996 on Taiwan. Write and solve an equation to find the year that the Sears Tower was completed.

15. FUNDRAISING During a five-day fundraiser, Shantell sold 8 boxes of greeting cards the first day, 6 boxes the second day, 10 boxes the third day, and 7 boxes the fourth day. If she sold a total of 45 boxes of greeting cards during the five days, write an equation that can be used to find the number of boxes Shantell sold the fifth day. Explain two methods of solving this equation. Then solve the equation.

16. ANALYZE TABLES The total points scored by both teams in the 2006 Super Bowl was 14 less than the total points for 2005. Write and solve an equation to find the total points for 2005.

Total Points Scored by Both Teams in Super Bowl	
Year	**Points**
2005	p
2006	31

1-10 Study Guide and Intervention

Solving Multiplication and Division Equations

You can use the following properties to solve multiplication and division equations.

- *Multiplication Property of Equality*— If you multiply each side of an equation by the same number, the two sides remain equal.
- *Division Property of Equality*— If you divide each side of an equation by the same nonzero number, the two sides remain equal.

Example 1 **Solve $19w = 114$. Check your solution.**

$19w = 114$	**Write the equation.**
$\frac{19w}{19} = \frac{114}{19}$	**Divide each side of the equation by 19.**
$1w = 6$	**$19 \div 19 = 1$ and $114 \div 19 = 6$.**
$w = 6$	**Identity Property; $1w = w$**

Check	$19w = 114$	**Write the original equation.**
	$19(6) \stackrel{?}{=} 114$	**Replace *w* with 6.**
	$114 = 114$ ✓	**This sentence is true.**

Example 2 **Solve $\frac{d}{15} = -9$. Check your solution.**

$$\frac{d}{15} = -9$$

$$\frac{d}{15}(15) = -9(15) \quad \text{Multiply each side of the equation by 15.}$$

$$d = -135$$

Check	$\frac{d}{15} = -9$	**Write the original equation.**
	$\frac{-135}{15} \stackrel{?}{=} -9$	**Replace *d* with −135.**
	$-9 = -9$ ✓	**$-135 \div 15 = -9$**

Exercises

Solve each equation. Check your solution.

1. $\frac{r}{5} = 6$ **2.** $2d = 12$ **3.** $7h = -21$

4. $-8x = 40$ **5.** $\frac{f}{8} = -6$ **6.** $\frac{x}{-10} = -7$

7. $17c = -68$ **8.** $\frac{h}{-11} = 12$ **9.** $29t = -145$

10. $125 = 5z$ **11.** $13t = -182$ **12.** $117 = -39k$

NAME ______________________________ DATE ____________ PERIOD _____

1-10 Practice

Solving Multiplication and Division Equations

Solve each equation. Check your solution.

1. $5s = 45$
2. $8h = 64$
3. $36 = 9b$
4. $-3p = 24$
5. $-12m = -72$
6. $-56 = 7d$
7. $\frac{x}{5} = 11$
8. $\frac{v}{4} = 20$
9. $\frac{c}{-2} = 43$
10. $16 = \frac{y}{-3}$
11. $-9 = \frac{n}{8}$
12. $\frac{a}{25} = -3$

13. **CARS** Mrs. Alvarez bought a new car. Her monthly payments are \$525. If she will pay a total of \$25,200 in payments, write and solve a multiplication equation to find the number of payments.

14. **POPULATION** The population of South Africa is four times the population of Greece. If the population of South Africa is 44 million, write and solve a multiplication equation to find the population of Greece.

MEASUREMENT For Exercises 15 and 16, refer to the table. Write and solve an equation to find each quantity.

15. the number of quarts in 24 pints

16. the number of gallons in 104 pints

Customary System Conversions (capacity)
1 pint = 2 cups
1 quart = 2 pints
1 quart = 4 cups
1 gallon = 4 quarts
1 gallon = 8 pints

Solve each equation.

17. $3 = \frac{-84}{g}$
18. $\frac{-4}{x} = -8$
19. $\frac{-144}{r} = -16$

NAME ______________________________ DATE ____________ PERIOD _____

2-1 Study Guide and Intervention

Rational Numbers

To express a fraction as a decimal, divide the numerator by the denominator.

Example 1 **Write $\frac{3}{4}$ as a decimal.**

$\frac{3}{4}$ means $3 \div 4$.

The fraction $\frac{3}{4}$ can be written as 0.75, since $3 \div 4 = 0.75$.

Example 2 **Write −0.16 as a fraction.**

$-0.16 = -\frac{16}{100}$ 0.16 is 16 hundredths.

$= -\frac{4}{25}$ Simplify.

The decimal −0.16 can be written as $-\frac{4}{25}$.

Example 3 **Write $8.\overline{2}$ as a mixed number.**

Let $N = 8.\overline{2}$ or 8.222… .

Then $10N = 82.222\ldots$.

Subtract.

$10N = 82.222\ldots$

$-1N = 8.222\ldots$ $N = 1N$

$9N = 74$ $10N - 1N = 9N$

$\frac{9N}{9} = \frac{74}{9}$ Divide each side by 9.

$N = 8\frac{2}{9}$ Simplify.

The decimal $8.\overline{2}$ can be written as $8\frac{2}{9}$.

Exercises **Write each fraction or mixed number as a decimal.**

1. $\frac{2}{5}$ **2.** $\frac{3}{10}$ **3.** $\frac{7}{8}$ **4.** $2\frac{16}{25}$

5. $-\frac{2}{3}$ **6.** $-1\frac{2}{9}$ **7.** $6\frac{2}{3}$ **8.** $-4\frac{3}{11}$

Write each decimal as a fraction or mixed number in simplest form.

9. 0.8 **10.** −0.15 **11.** $0.\overline{1}$ **12.** $1.\overline{7}$

NAME ______________________ DATE ____________ PERIOD _____

2-1 Practice

Rational Numbers

Write each fraction or mixed number as a decimal.

1. $\frac{3}{5}$ **2.** $\frac{5}{8}$ **3.** $\frac{9}{20}$

4. $\frac{37}{50}$ **5.** $-\frac{11}{16}$ **6.** $-\frac{9}{32}$

7. $3\frac{1}{5}$ **8.** $4\frac{3}{8}$ **9.** $\frac{5}{33}$

10. $-\frac{7}{9}$ **11.** $-8\frac{11}{18}$ **12.** $-9\frac{11}{30}$

Write each decimal as a fraction or mixed number in simplest form.

13. -0.8 **14.** 0.44 **15.** 1.35

16. $0.\overline{8}$ **17.** $-1.\overline{5}$ **18.** $4.\overline{4}$

POPULATION **For Exercises 19–21, refer to the table at the right.**

19. Express the fraction for Asian as a decimal.

20. Find the decimal equivalent for the fraction of the population that is African American.

21. Write the fraction for Hispanic as a decimal. Round to the nearest thousandth.

Population of California by Race

Race	Fraction of Total Population
Asian	$\frac{1}{10}$
African American	$\frac{1}{16}$
Hispanic	$\frac{1}{3}$

Source: U.S. Census Bureau

MEASUREMENTS **For Exercises 22 and 23, use the figure at the right.**

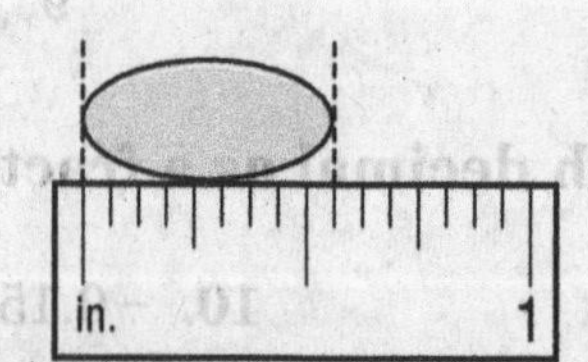

22. Write the width of the jellybean as a fraction.

23. Write the width of the jellybean as a decimal.

2-2 Study Guide and Intervention

Comparing and Ordering Rational Numbers

When comparing two or more rational numbers, either write the numbers as fractions with the same denominator or write the numbers as decimals.

Example 1 **Replace ● with <, >, or = to make $\frac{4}{5}$ ● $\frac{7}{10}$ a true sentence.**

Write as fractions with the same denominator. The least common denominator is 10.

$\frac{4}{5} = \frac{4 \cdot 2}{5 \cdot 2}$ or $\frac{8}{10}$

$\frac{7}{10} = \frac{7 \cdot 1}{10 \cdot 1}$ or $\frac{7}{10}$

Since $\frac{8}{10} > \frac{7}{10}$, $\frac{4}{5} > \frac{7}{10}$.

Example 2 **Order the set of rational numbers -3.25, $-3\frac{1}{3}$, $-3\frac{2}{5}$, and $-3.2\overline{5}$ from least to greatest.**

Write $-3\frac{1}{3}$ and $-3\frac{2}{5}$ as decimals.

$\frac{1}{3} = 0.\overline{3}$, so $-3\frac{1}{3} = -3.\overline{3}$.

$\frac{2}{5} = 0.4$, so $-3\frac{2}{5} = -3.4$.

Since $-3.4 < -3.\overline{3} < -3.2\overline{5} < -3.25$, the numbers from least to greatest are $-3\frac{2}{5}$, $-3\frac{1}{3}$, $-3.2\overline{5}$, and -3.25.

Exercises

Replace each ● with <, >, or = to make a true sentence.

1. $\frac{5}{6}$ ● $\frac{2}{3}$
2. $\frac{4}{5}$ ● $\frac{13}{15}$
3. $\frac{1}{9}$ ● $\frac{1}{8}$
4. $-\frac{2}{3}$ ● $-\frac{7}{10}$
5. $3\frac{7}{10}$ ● $3\frac{4}{5}$
6. $-2\frac{3}{7}$ ● $-2\frac{4}{9}$
7. 2.6 ● $2\frac{5}{8}$
8. $4\frac{1}{6}$ ● $4.1\overline{6}$
9. $-4.5\overline{8}$ ● $-4.\overline{58}$

Order each set of rational numbers from least to greatest.

10. 0.5, 0.1, $\frac{1}{4}$, $\frac{2}{3}$
11. 2.4, $2\frac{4}{7}$, 2.13, $1\frac{9}{10}$
12. $\frac{1}{5}$, -0.7, 0.25, $-\frac{3}{5}$
13. $1\frac{2}{9}$, $1\frac{2}{3}$, 1.45, 1.67
14. $-2\frac{1}{4}$, -2.28, -2.7, $-2\frac{4}{5}$
15. $4\frac{2}{3}$, $4\frac{5}{6}$, 4.6, 5.3

NAME ______________________________ DATE ____________ PERIOD _____

2-2 Practice

Comparing and Ordering Rational Numbers

Replace each ● with <, >, or = to make a true sentence.

1. $\frac{3}{5}$ ● $\frac{5}{7}$ 2. $\frac{4}{9}$ ● $\frac{5}{11}$ 3. $3\frac{2}{11}$ ● $3\frac{1}{9}$ 4. $5\frac{7}{15}$ ● $5\frac{8}{17}$

5. 0.2 ● $\frac{2}{11}$ 6. 0.25 ● $\frac{5}{21}$ 7. $8\frac{10}{27}$ ● 8.3 8. $4\frac{8}{30}$ ● 4.3

9. $-\frac{8}{13}$ ● $-\frac{5}{13}$ 10. $-\frac{3}{8}$ ● $-\frac{7}{8}$ 11. $-\frac{2}{5}$ ● $-\frac{6}{7}$ 12. $-\frac{2}{9}$ ● $-\frac{9}{11}$

13. −4.5 ● −4.55 14. −6.14 ● −6.15 15. −3.57 ● −3.5 16. −1.9 ● −1.99

17. Which is least: $\frac{3}{8}$, 0.4, $\frac{4}{11}$, $0.03\overline{5}$, or $\frac{5}{13}$?

18. Which is greatest: $\frac{7}{9}$, 0.778, $0.\overline{78}$, $\frac{11}{13}$, or 0.787?

Order each set of rational numbers from least to greatest.

19. −5.81, $-5\frac{3}{4}$, $-5\frac{3}{5}$, −5.69

20. −1.01, −1.1, $-1\frac{1}{9}$, $-1\frac{1}{11}$

21. Which point on the number line is the graph of 0.875?

P Q S R
0 $\frac{1}{4}$ $\frac{1}{2}$ $\frac{3}{4}$ 1

22. **STATISTICS** If you order a set of numbers from least to greatest, the middle number is the *median*. Find the median of 43.7, 41.3, 44.5, $42\frac{4}{5}$, and $43\frac{3}{4}$.

NAME ______________________ DATE __________ PERIOD _____

2-3 Study Guide and Intervention

Multiplying Positive and Negative Fractions

To multiply fractions, multiply the numerators and multiply the denominators.

Example 1 **Find $\frac{3}{8} \cdot \frac{4}{11}$. Write in simplest form.**

$\frac{3}{8} \cdot \frac{4}{11} = \frac{3}{\cancel{8}_2} \cdot \frac{\cancel{4}^1}{11}$ Divide 8 and 4 by their GCF, 4.

$= \frac{3 \cdot 1}{2 \cdot 11}$ Multiply the numerators and denominators.

$= \frac{3}{22}$ Simplify.

To multiply mixed numbers, first rewrite them as improper fractions.

Example 2 **Find $-2\frac{1}{3} \cdot 3\frac{3}{5}$. Write in simplest form.**

$-2\frac{1}{3} \cdot 3\frac{3}{5} = -\frac{7}{3} \cdot \frac{18}{5}$ $-2\frac{1}{3} = -\frac{7}{3}, 3\frac{3}{5} = \frac{18}{5}$

$= -\frac{7}{\cancel{3}_1} \cdot \frac{\cancel{18}^6}{5}$ Divide 18 and 3 by their GCF, 3.

$= -\frac{7 \cdot 6}{1 \cdot 5}$ Multiply the numerators and denominators.

$= -\frac{42}{5}$ Simplify.

$= -8\frac{2}{5}$ Write the result as a mixed number.

Exercises

Multiply. Write in simplest form.

1. $\frac{2}{3} \cdot \frac{3}{5}$ **2.** $\frac{4}{7} \cdot \frac{3}{4}$ **3.** $-\frac{1}{2} \cdot \frac{7}{9}$

4. $\frac{9}{10} \cdot \frac{2}{3}$ **5.** $\frac{5}{8} \cdot \left(-\frac{4}{9}\right)$ **6.** $-\frac{4}{7} \cdot \left(-\frac{2}{3}\right)$

7. $2\frac{2}{5} \cdot \frac{1}{6}$ **8.** $-3\frac{1}{3} \cdot 1\frac{1}{2}$ **9.** $3\frac{3}{7} \cdot 2\frac{5}{8}$

10. $-1\frac{7}{8} \cdot \left(-2\frac{2}{5}\right)$ **11.** $-1\frac{3}{4} \cdot 2\frac{1}{5}$ **12.** $2\frac{2}{3} \cdot 2\frac{3}{7}$

NAME ______________________ DATE ____________ PERIOD _____

2-3 Practice

Multiplying Positive and Negative Fractions

Find each product. Write in simplest form.

1. $\frac{1}{4} \cdot \frac{4}{5}$

2. $\frac{6}{7} \cdot \frac{1}{2}$

3. $\frac{3}{10} \cdot \frac{2}{3}$

4. $-\frac{15}{16} \cdot \frac{4}{5}$

5. $\left(-\frac{8}{25}\right)\frac{15}{16}$

6. $\left(-\frac{7}{8}\right)\left(-\frac{1}{7}\right)$

7. $1\frac{1}{4} \cdot \frac{1}{5}$

8. $1\frac{1}{4} \cdot 1\frac{1}{5}$

9. $-2\frac{2}{3} \cdot \left(-\frac{1}{4}\right)$

10. $\frac{1}{4} \cdot \left(-\frac{4}{15}\right) \cdot \frac{5}{7}$

11. $2\frac{2}{5} \cdot 2\frac{1}{3} \cdot 2$

12. $10 \cdot 8.56 \cdot \frac{1}{2}$

ALGEBRA Evaluate each expression if $a = -\frac{1}{5}$, $b = \frac{2}{3}$, $c = \frac{7}{8}$, and $d = -\frac{3}{4}$.

13. bc

14. ab

15. abc

16. abd

17. **COOKING** A recipe calls for $2\frac{1}{4}$ cups of flour. How much flour would you need to make $\frac{1}{3}$ of the recipe?

18. **FARMING** A farmer has $6\frac{1}{2}$ acres of land for growing crops. If she plants corn on $\frac{3}{5}$ of the land, how many acres of corn will she have?

ALGEBRA Evaluate each expression if $e = -1\frac{1}{4}$, $f = 2\frac{2}{3}$, $g = -2\frac{1}{6}$, and $h = 1\frac{1}{5}$.

19. efh^2

20. e^2h^2

21. $\frac{1}{8}f^2g$

22. $-2ef(-gh)$

NAME ______________________ DATE ____________ PERIOD ______

2-4 Study Guide and Intervention

Dividing Positive and Negative Fractions

Two numbers whose product is 1 are **multiplicative inverses**, or **reciprocals**, of each other.

Example 1 **Write the multiplicative inverse of $-2\frac{3}{4}$.**

$-2\frac{3}{4} = -\frac{11}{4}$ Write $-2\frac{3}{4}$ as an improper fraction.

Since $-\frac{11}{4}\left(-\frac{4}{11}\right) = 1$, the multiplicative inverse of $-2\frac{3}{4}$ is $-\frac{4}{11}$.

To divide by a fraction or mixed number, multiply by its multiplicative inverse.

Example 2 **Find $\frac{3}{8} \div \frac{6}{7}$. Write in simplest form.**

$\frac{3}{8} \div \frac{6}{7} = \frac{3}{8} \cdot \frac{7}{6}$ Multiply by the multiplicative inverse of $\frac{6}{7}$, which is $\frac{7}{6}$.

$= \frac{\overset{1}{\cancel{3}}}{8} \cdot \frac{7}{\underset{2}{\cancel{6}}}$ Divide 6 and 3 by their GCF, 3.

$= \frac{7}{16}$ Simplify.

Exercises

Write the multiplicative inverse of each number.

1. $\frac{3}{5}$ **2.** $-\frac{8}{9}$ **3.** $\frac{1}{10}$ **4.** $-\frac{1}{6}$

5. $2\frac{3}{5}$ **6.** $-1\frac{2}{3}$ **7.** $-5\frac{2}{5}$ **8.** $7\frac{1}{4}$

Divide. Write in simplest form.

9. $\frac{1}{3} \div \frac{1}{6}$ **10.** $\frac{2}{5} \div \frac{4}{7}$

11. $-\frac{5}{6} \div \frac{3}{4}$ **12.** $1\frac{1}{5} \div 2\frac{1}{4}$

13. $3\frac{1}{7} \div \left(-3\frac{2}{3}\right)$ **14.** $-\frac{4}{9} \div 2$

15. $\frac{6}{11} \div (-4)$ **16.** $5 \div 2\frac{1}{3}$

Lesson 2-4

NAME ______________________________ DATE __________ PERIOD _____

2-4 Practice

Dividing Positive and Negative Fractions

Write the multiplicative inverse of each number.

1. $\frac{4}{5}$ **2.** $\frac{7}{12}$ **3.** -20 **4.** $-5\frac{3}{8}$

Find each quotient. Write in simplest form.

5. $\frac{1}{5} \div \frac{1}{4}$ **6.** $\frac{2}{5} \div \frac{5}{6}$ **7.** $\frac{3}{7} \div \frac{6}{11}$

8. $\frac{3}{10} \div \frac{4}{5}$ **9.** $\frac{3}{8} \div 6$ **10.** $\frac{6}{7} \div 3$

11. $\frac{4}{5} \div 10$ **12.** $\frac{6}{11} \div 8$ **13.** $-\frac{4}{5} \div \frac{5}{6}$

14. $\frac{5}{12} \div \left(-\frac{3}{5}\right)$ **15.** $-\frac{3}{10} \div \left(-\frac{2}{5}\right)$ **16.** $-\frac{13}{18} \div \left(-\frac{8}{9}\right)$

17. $4\frac{1}{5} \div 1\frac{3}{4}$ **18.** $8\frac{1}{3} \div 3\frac{3}{4}$ **19.** $-10\frac{1}{2} \div 2\frac{1}{3}$

20. **OFFICE SUPPLIES** A regular paper clip is $1\frac{1}{4}$ inches long, and a jumbo paper clip is $1\frac{7}{8}$ inches long. How many times longer is the jumbo paper clip than the regular paper clip?

21. **STORAGE** The ceiling in a storage unit is $7\frac{2}{3}$ feet high. How many boxes may be stacked in a single stack if each box is $\frac{3}{4}$ foot tall?

ALGEBRA Evaluate each expression for the given values.

22. $r \div s$ if $r = -\frac{7}{20}$ and $s = \frac{7}{15}$ **23.** $m \div n$ if $m = \frac{4}{9}$ and $n = \frac{11}{12}$

2-5 Study Guide and Intervention

Adding and Subtracting Like Fractions

Fractions that have the same denominator are called **like fractions**. To add like fractions, add the numerators of the fractions and write the sum over the denominator.

Example 1 **Find $\frac{1}{5} + \left(-\frac{4}{5}\right)$. Write in simplest form.**

$\frac{1}{5} + \left(-\frac{4}{5}\right) = \frac{1 + (-4)}{5}$ Add the numerators. The denominators are the same.

$= \frac{-3}{5}$ or $-\frac{3}{5}$ Simplify.

To subtract like fractions, subtract the numerators of the fractions and write the difference over the denominator.

Example 2 **Find $-\frac{4}{9} - \frac{7}{9}$. Write in simplest form.**

$-\frac{4}{9} - \frac{7}{9} = \frac{-4 - 7}{9}$ Subtract the numerators. The denominators are the same.

$= \frac{-11}{9}$ or $-1\frac{2}{9}$ Rename $\frac{-11}{9}$ as $-1\frac{2}{9}$.

To add or subtract mixed numbers, first write the mixed numbers as improper fractions. Then add or subtract the improper fractions and simplify the result.

Example 3 **Find $2\frac{3}{7} + 6\frac{5}{7}$. Write in simplest form.**

$2\frac{3}{7} + 6\frac{5}{7} = \frac{17}{7} + \frac{47}{7}$ Write the mixed numbers as improper fractions.

$= \frac{17 + 47}{7}$ Add the numerators. The denominators are the same.

$= \frac{64}{7}$ or $9\frac{1}{7}$ Rewrite $\frac{64}{7}$ as $9\frac{1}{7}$.

Exercises

Add or subtract. Write in simplest form.

1. $\frac{4}{7} + \frac{2}{7}$
2. $\frac{1}{10} + \frac{5}{10}$
3. $\frac{5}{9} + \left(-\frac{1}{9}\right)$
4. $\frac{1}{6} + \left(-\frac{5}{6}\right)$
5. $-\frac{3}{8} + \frac{7}{8}$
6. $\frac{5}{11} - \left(-\frac{4}{11}\right)$
7. $-\frac{4}{5} - \frac{3}{5}$
8. $-\frac{9}{13} + \left(-\frac{6}{13}\right)$
9. $2\frac{1}{4} + 1\frac{1}{4}$
10. $3\frac{5}{7} + 2\frac{3}{7}$
11. $3\frac{5}{8} - 1\frac{3}{8}$
12. $4\frac{3}{5} - 2\frac{4}{5}$

NAME ______________________________ DATE ____________ PERIOD _____

2-5 Practice

Adding and Subtracting Like Fractions

Add or subtract. Write in simplest form.

1. $-\frac{1}{4} + \frac{3}{4}$

2. $-\frac{3}{8} + \left(-\frac{1}{8}\right)$

3. $-\frac{8}{11} + \frac{10}{11}$

4. $-\frac{5}{7} - \frac{4}{7}$

5. $\frac{11}{12} - \frac{7}{12}$

6. $\frac{2}{15} + \frac{7}{15}$

7. $4\frac{3}{4} + 6\frac{3}{4}$

8. $5\frac{7}{10} + 9\frac{9}{10}$

9. $7\frac{4}{9} + \left(-3\frac{5}{9}\right)$

10. $-1\frac{8}{9} - 4\frac{8}{9}$

11. $-4\frac{4}{5} - 5\frac{4}{5}$

12. $8\frac{5}{6} - 3\frac{5}{6}$

13. SEWING Naomi needs $2\frac{3}{4}$ yards of fabric to make a banner for a football game. The fabric store has $6\frac{1}{4}$ yards of the fabric she wants. How much of the fabric will remain at the store after Naomi buys her fabric?

14. GEOMETRY Find the perimeter of the triangle.

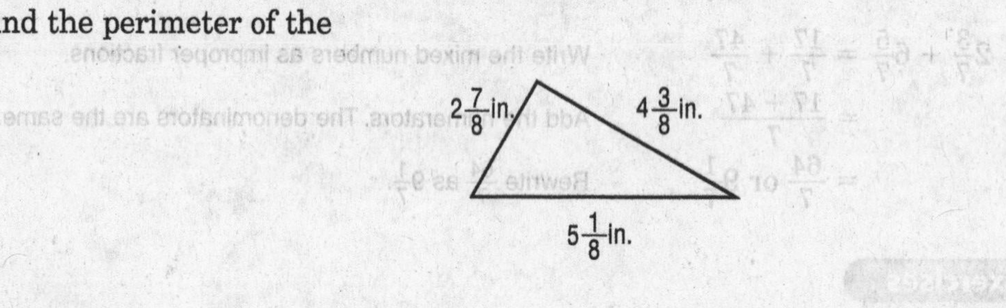

Simplify each expression.

15. $-5\frac{4}{7} + 2\frac{1}{7} - \left(3\frac{5}{7}\right)$

16. $-7\frac{1}{12} - \left(-4\frac{11}{12}\right) + 9\frac{7}{12}$

ALGEBRA Evaluate each expressions for the given values.

17. $r + s$ if $r = 8\frac{4}{5}$ and $s = -3\frac{2}{5}$

18. $b - c$ if $b = -2\frac{7}{9}$ and $c = -9\frac{5}{9}$

NAME ______________________ DATE ____________ PERIOD _____

2-6 Study Guide and Intervention

Adding and Subtracting Unlike Fractions

Fractions with unlike denominators are called **unlike fractions.** To add or subtract unlike fractions, rename the fractions using the least common denominator. Then add or subtract as with like fractions.

Example 1 **Find $\frac{3}{5} + \frac{2}{3}$. Write in simplest form.**

$\frac{3}{5} + \frac{2}{3} = \frac{3}{5} \cdot \frac{3}{3} + \frac{2}{3} \cdot \frac{5}{5}$ — The LCD is 5 · 3 or 15.

$= \frac{9}{15} + \frac{10}{15}$ — Rename each fraction using the LCD.

$= \frac{9 + 10}{15}$ — Add the numerators. The denominators are the same.

$= \frac{19}{15}$ or $1\frac{4}{15}$ — Simplify.

Example 2 **Find $-3\frac{1}{2} - 1\frac{5}{6}$. Write in simplest form.**

$-3\frac{1}{2} - 1\frac{5}{6} = -\frac{7}{2} - \frac{11}{6}$ — Write the mixed numbers as improper fractions.

$= -\frac{7}{2} \cdot \frac{3}{3} - \frac{11}{6}$ — The LCD is 2 · 3 or 6.

$= -\frac{21}{6} - \frac{11}{6}$ — Rename $\frac{7}{2}$ using the LCD.

$= \frac{-21 - 11}{6}$ — Subtract the numerators.

$= -\frac{32}{6}$ or $-\frac{16}{3}$ or $-5\frac{1}{3}$ — Simplify.

Exxercises

Add or subtract. Write in simplest form.

1. $\frac{2}{5} + \frac{3}{10}$

2. $\frac{1}{3} + \frac{2}{9}$

3. $\frac{5}{9} + \left(-\frac{1}{6}\right)$

4. $-\frac{3}{4} - \frac{5}{6}$

5. $\frac{4}{5} - \left(-\frac{1}{3}\right)$

6. $1\frac{2}{3} - \left(-\frac{4}{9}\right)$

7. $-\frac{7}{10} - \left(-\frac{1}{2}\right)$

8. $2\frac{1}{4} + 1\frac{3}{8}$

9. $3\frac{3}{4} - 1\frac{1}{3}$

10. $-1\frac{1}{5} - 2\frac{1}{4}$

11. $-2\frac{4}{9} - \left(-1\frac{1}{3}\right)$

12. $3\frac{3}{5} - 2\frac{2}{3}$

NAME ______________________ DATE ____________ PERIOD _____

2-6 Practice

Adding and Subtracting Unlike Fractions

Add or subtract. Write in simplest form.

1. $-\frac{1}{2} + \frac{7}{10}$

2. $\frac{5}{6} + \left(-\frac{5}{9}\right)$

3. $-\frac{4}{5} + \left(-\frac{1}{3}\right)$

4. $\frac{7}{9} - \frac{2}{5}$

5. $\frac{3}{4} - \left(-\frac{1}{12}\right)$

6. $-\frac{7}{8} - \left(-\frac{2}{3}\right)$

7. $4\frac{1}{5} + 6\frac{3}{4}$

8. $1\frac{7}{10} + \left(-5\frac{3}{5}\right)$

9. $7\frac{3}{5} - \left(-5\frac{1}{3}\right)$

10. $-3\frac{2}{3} - 4\frac{5}{9}$

11. $-4\frac{3}{5} - 5\frac{9}{10}$

12. $-18\frac{5}{12} + 14\frac{3}{4}$

13. POPULATION About $\frac{1}{5}$ of the world's population lives in China, and $\frac{1}{6}$ of the world's population lives in India. What fraction of the world's population lives in other countries?

ALGEBRA For Exercises 14 and 15, evaluate each expression using the given information.

14. $m - n$ if $m = -\frac{3}{5}$ and $n = -10\frac{7}{10}$

15. $j - k$ if $j = -\frac{5}{9}$ and $k = 4\frac{5}{6}$

GEOMETRY **Find the missing measure for each figure.**

16.

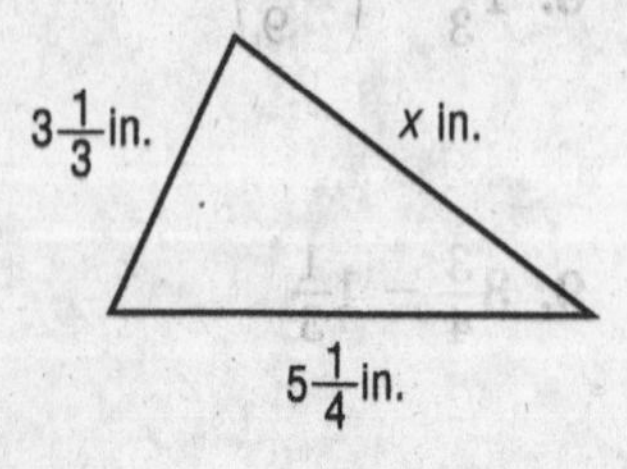

perimeter = $12\frac{23}{24}$ in.

17.

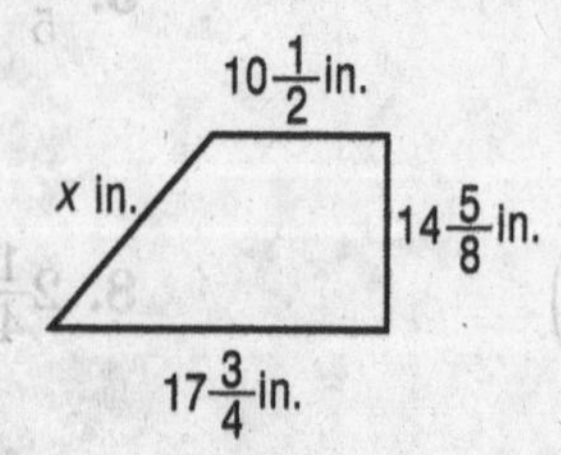

perimeter = $59\frac{1}{4}$ in.

2-7 Study Guide and Intervention

Solving Equations with Rational Numbers

The Addition, Subtraction, Multiplication, and Division Properties of Equality can be used to solve equations with rational numbers.

Example 1 **Solve $x - 2.73 = 1.31$. Check your solution.**

	$x - 2.73 = 1.31$	Write the equation.
	$x - 2.73 + 2.73 = 1.31 + 2.73$	Add 2.73 to each side.
	$x = 4.04$	Simplify.
Check	$x - 2.73 = 1.31$	Write the original equation.
	$4.04 - 2.73 \stackrel{?}{=} 1.31$	Replace x with 4.04.
	$1.31 = 1.31$ ✓	Simplify.

Example 2 **Solve $\frac{4}{5}y = \frac{2}{3}$. Check your solution.**

	$\frac{4}{5}y = \frac{2}{3}$	Write the equation.
	$\frac{5}{4}\left(\frac{4}{5}y\right) = \frac{5}{4} \cdot \frac{2}{3}$	Multiply each side by $\frac{5}{4}$.
	$y = \frac{5}{6}$	Simplify.
Check	$\frac{4}{5}y = \frac{2}{3}$	Write the original equation.
	$\frac{4}{5}\left(\frac{5}{6}\right) \stackrel{?}{=} \frac{2}{3}$	Replace y with $\frac{5}{6}$.
	$\frac{2}{3} = \frac{2}{3}$ ✓	Simplify.

Exercises

Solve each equation. Check your solution.

1. $t + 1.32 = 3.48$

2. $b - 4.22 = 7.08$

3. $-8.07 = r - 4.48$

4. $h + \frac{4}{9} = \frac{7}{9}$

5. $-\frac{5}{8} = x - \frac{1}{4}$

6. $-\frac{2}{3} + f = \frac{3}{5}$

7. $3.2c = 9.6$

8. $-5.04 = 1.26d$

9. $\frac{3}{5}x = 6$

10. $-\frac{2}{3} = \frac{3}{4}t$

11. $\frac{w}{2.5} = 4.2$

12. $1\frac{3}{4}r = 3\frac{5}{8}$

NAME ______________________________ DATE ____________ PERIOD _____

2-7 Practice

Solving Equations with Rational Numbers

Solve each equation. Check your solution.

1. $m + 0.88 = 1.64$

2. $t - 2.89 = 9.15$

3. $-\frac{3}{5} = d - \frac{5}{6}$

4. $-\frac{7}{16} = b + \frac{1}{4}$

5. $h - (-6.3) = 8.12$

6. $-2.5 = n - (-5.37)$

7. $-\frac{5}{8}k = 25$

8. $-\frac{3}{7}v = -27$

9. $-2.94 = -0.42a$

10. $-8.4 = 1.4y$

11. $\frac{f}{2.4} = -7.5$

12. $\frac{p}{-6.25} = -3.6$

13. $2.5x = -\frac{13}{16}$

14. $-4.5w = -8\frac{1}{3}$

15. $8\frac{2}{3} = -1.\overline{3}g$

17. MONEY The currency in Switzerland is called a franc. On a certain day, one U.S. dollar equaled $1\frac{1}{4}$ Swiss francs. Write and solve a multiplication equation to find the number of U.S. dollars that would equal 15 Swiss francs.

FOOTBALL For Exercise 18, refer to the table.

18. Let s equal the number of additional seats that the Pittsburgh Steelers' stadium needs to equal the number of seats in Kansas City Chiefs' stadium. Write and solve an addition equation to determine the number of seats that the Steelers' stadium needs to equal the number of seats in the Chiefs' stadium.

NFL Stadiums Seating Capacity

Stadium	Seats (thousands)
Dallas Cowboys	65.7
Kansas City Chiefs	79.4
Pittsburgh Steelers	64.5
San Diego Chargers	71.3

2-8 Study Guide and Intervention

Problem-Solving Investigation: Look for a Pattern

You may need to look for a pattern to solve a problem.

Understand	Determine what information is given in the problem and what you need to find.
Plan	Select a strategy including a possible estimate.
Solve	Solve the problem by carrying out your plan.
Check	Examine your answer to see if it seems reasonable.

Example

Three people board the subway train at the first stop. Five people board the train at the second stop. Seven people board the train at the third stop. If this pattern continues and no one gets off the train, how many people are on the subway train when it reaches the seventh and final stop?

Understand You know that 3 people boarded the subway train at the first stop. At each subsequent stop, 2 more people board the train than at the previous stop.

Plan Look for a pattern and use the pattern to find how many people boarded the train in all.

Solve Complete the information for the first, second, and third stops. Continue the pattern to solve the problem.

First Stop	Second Stop	Third Stop	Fourth Stop	Fifth Stop	Sixth Stop	Seventh Stop
3	5	7	9	11	13	15
3 people on the train	3 + 5 = 8 people on the train	8 + 7 = 15 people on the train	15 + 9 = 24 people on the train	24 + 11 = 35 people on the train	35 + 13 = 48 people on the train	48 + 15 = 63 people on the train

At the seventh and final stop there were 63 people on the subway train.

Check Check your pattern to make sure the answer is correct.

Exercises

Look for a pattern. Then use the pattern to solve each problem.

1. **COOKING** A muffin recipe calls for $2\frac{1}{2}$ cups of flour for every $\frac{2}{3}$ cup of sugar. How many cups of flour should be used when 4 cups of sugar are used?

2. **FUNDRAISER** There were 256 people at a fundraiser. When the event was over, half of the people who remained left every 5 minutes. How long after the event ended did the last person leave?

NAME ______________________ DATE __________ PERIOD _____

2-8 Practice

Problem-Solving Investigation: Look for a Pattern

Mixed Problem Solving

For Exercises 1 and 2, look for a pattern. Then use the pattern to solve the problem.

1. **GEOMETRY** Draw the next two angles in the pattern.

a. 10°

b. 20°

c. 30°

d. 40°

2. **ANALYZE TABLES** A falling object continues to fall faster until it hits the ground. How far will an object fall in five seconds?

Time Period	Distance Fallen
1 second	16 feet
2 seconds	48 feet
3 seconds	80 feet
4 seconds	112 feet

Use any strategy to solve Exercises 3–6. Some strategies are shown below.

PROBLEM-SOLVING STRATEGIES
• Look for a pattern.
• Work backward.

3. **YARD WORK** Denzel can mow $\frac{1}{8}$ of his yard every 7 minutes. If he has 40 minutes to mow $\frac{3}{4}$ of the yard, will he have enough time?

4. **READING** Ling read 175 pages by 1:00 P.M., 210 pages by 2:00 P.M., and 245 pages by 3:00 P.M. If she continues reading at this rate, how many pages will Ling have read by 4:00 P.M.?

5. **MOVIES** The land area of Alaska is about 570 thousand square miles. The land area of Washington, D.C., is about 0.06 thousand square miles. How many times larger is Alaska than Washington, D.C.?

6. **U.S. PRESIDENTS** President Clinton served 5 two-year terms as governor of Arkansas and 2 four-year terms as President of the United States. How many total years did he serve in these two government offices?

NAME ______________________________ DATE ____________ PERIOD _____

2-9 Study Guide and Intervention

Powers and Exponents

Expressions containing repeated factors can be written using exponents.

Example 1 **Write $7 \cdot 7 \cdot 7 \cdot 7 \cdot 7$ using exponents.**

Since 7 is used as a factor 5 times, $7 \cdot 7 \cdot 7 \cdot 7 \cdot 7 = 7^5$.

Example 2 **Write $p \cdot p \cdot p \cdot q \cdot q$ using exponents.**

Since p is used as a factor 3 times and q is used as a factor 2 times, $p \cdot p \cdot p \cdot q \cdot q = p^3 \cdot q^2$.

Any nonzero number to the zero power is 1. Any nonzero number to the negative *n* power is the multiplicative inverse of the *n*th power.

Example 3 **Evaluate 6^2.**

$6^2 = 6 \cdot 6$ Definition of exponents

$= 36$ Simplify.

Example 4 **Evaluate 5^{-3}.**

$5^{-3} = \frac{1}{5^3}$ Definition of negative exponents

$= \frac{1}{125}$ Simplify.

Exercises

Write each expression using exponents.

1. $8 \cdot 8 \cdot 8 \cdot 8 \cdot 8$

2. $4 \cdot 4 \cdot 4 \cdot 4$

3. $a \cdot a \cdot a \cdot a \cdot a \cdot a$

4. $g \cdot g \cdot g \cdot g \cdot g \cdot g \cdot g$

5. $5 \cdot 5 \cdot 9 \cdot 9 \cdot 5 \cdot 9 \cdot 5 \cdot 5$

6. $s \cdot w \cdot w \cdot s \cdot s \cdot s$

Evaluate each expression.

7. 4^2

8. 5^3

9. 13^2

10. $2^3 \cdot 3^2$

11. 8^{-2}

12. $2^4 \cdot 5^2$

13. 3^{-4}

14. $3^4 \cdot 7^2$

Lesson 2-9

NAME ______________________ DATE __________ PERIOD _____

2-9 Practice

Powers and Exponents

Write each expression using exponents.

1. $3 \cdot 3 \cdot m$

2. $2 \cdot d \cdot 5 \cdot d \cdot d \cdot 5$

3. $p \cdot 9 \cdot 3 \cdot q \cdot p \cdot 9$

4. $g \cdot 7 \cdot 7 \cdot g \cdot h \cdot 7 \cdot h$

5. $2 \cdot 5 \cdot r \cdot 7 \cdot s \cdot r \cdot 5 \cdot r \cdot 7 \cdot r \cdot s$

6. $x \cdot 8 \cdot y \cdot x \cdot 5 \cdot x \cdot 5 \cdot y \cdot 8 \cdot y \cdot y \cdot 5$

Evaluate each expression.

7. 2^4

8. 5^3

9. $2^2 \cdot 6^2$

10. $2^3 \cdot 5^2$

11. 3^{-4}

12. 8^{-3}

13. 9^{-2}

14. 5^{-3}

15. $7 \cdot 2^2 \cdot 5^2$

16. $3^2 \cdot 6 \cdot 10^2$

17. $3^{-2} \cdot 2^{-3}$

18. $7 \cdot 3^3 \cdot 5^{-4}$

ALGEBRA Evaluate each expression.

19. $r^3 \cdot s$, if $r = 5$ and $s = 4$

20. $m^2 \cdot n^3$, if $m = 6$ and $n = 2$

21. $f^4 \cdot g^5$, if $f = 3$ and $g = 1$

22. $x^5 \cdot y$, if $x = 2$ and $y = 8$

23. Complete the following pattern.
$5^4 = 625$, $5^3 = 125$, $5^2 = 25$, $5^1 = 5$, $5^0 = \underline{?}$, $5^{-1} = \underline{?}$, $5^{-2} = \underline{?}$, $5^{-3} = \underline{?}$

24. MONEY Suppose $100 is deposited into an account and the amount doubles every 8 years. How much will be in the account after 40 years?

25. EPIDEMICS At the beginning of an epidemic, 50 people are sick. If the number of sick people triples every other day, how many people will be sick at the end of 2 weeks?

NAME ______________________ DATE __________ PERIOD _____

2-10 Study Guide and Intervention

Scientific Notation

A number in scientific notation is written as the product of a factor that is at least one but less than ten and a power of ten.

Example 1 **Write 8.65×10^7 in standard form.**

$8.65 \times 10^7 = 8.65 \times 10{,}000{,}000$ — $10^7 = 10 \cdot 10 \cdot 10 \cdot 10 \cdot 10 \cdot 10 \cdot 10$ or 10,000,000

$= 86{,}500{,}000$ — Move the decimal point 7 places to the right.

Example 2 **Write 9.2×10^{-3} in standard form.**

$9.2 \times 10^{-3} = 9.2 \times \frac{1}{10^3}$ — $10^{-3} = \frac{1}{10^3}$

$= 9.2 \times 0.001$ — $\frac{1}{10^3} = \frac{1}{1{,}000}$ or 0.001

$= 0.0092$ — Move the decimal point 3 places to the left.

Example 3 **Write 76,250 in scientific notation.**

$76{,}250 = 7.625 \times 10{,}000$ — The decimal point moves 4 places.

$= 7.625 \times 10^4$ — The exponent is positive.

Example 4 **Write 0.00157 in scientific notation.**

$0.00157 = 1.57 \times 0.001$ — The decimal point moves 3 places.

$= 1.57 \times 10^{-3}$ — The exponent is negative.

Exercises

Write each number in standard form.

1. 5.3×10^1

2. 9.4×10^3

3. 7.07×10^5

4. 2.6×10^{-3}

5. 8.651×10^{-2}

6. 6.7×10^{-6}

Write each number in scientific notation.

7. 561

8. 14

9. 56,400,000

10. 0.752

11. 0.0064

12. 0.000581

Lesson 2-10

NAME ______________________ DATE __________ PERIOD _____

2-10 Practice

Scientific Notation

Write each number in standard form.

1. 9.03×10^2 **2.** 7.89×10^3 **3.** 4.115×10^5 **4.** 3.201×10^6

5. 5.1×10^{-2} **6.** 7.7×10^{-5} **7.** 3.85×10^{-4} **8.** 1.04×10^{-3}

Write each number in scientific notation.

9. 4,400 **10.** 75,000 **11.** 69,900,000 **12.** 575,000,000

13. 0.084 **14.** 0.0099 **15.** 0.000000515 **16.** 0.0000307

17. Which number is greater: 3.5×10^4 or 2.1×10^6?

18. Which number is less: 7.2×10^7 or 9.9×10^5?

19. **POPULATION** The table lists the populations of five countries. List the countries from least to greatest population.

Country	Population
Australia	2.0×10^7
Brazil	1.9×10^8
Egypt	7.7×10^7
Luxembourg	4.7×10^5
Singapore	4.4×10^6

Source: *The World Factbook*

20. **SOLAR SYSTEM** Pluto is 3.67×10^9 miles from the Sun. Write this number in standard form.

21. **MEASUREMENT** One centimeter is equal to about 0.0000062 mile. Write this number in scientific notation.

22. **DISASTERS** In 2005, Hurricane Katrina caused over $125 billion in damage in the southern United States. Write $125 billion in scientific notation.

3-1 Study Guide and Intervention

Square Roots

The square root of a number is one of two equal factors. The radical sign $\sqrt{\ }$ is used to indicate a square root.

Examples **Find each square root.**

1. $\sqrt{1}$ Since $1 \cdot 1 = 1$, $\sqrt{1} = 1$.

2. $-\sqrt{16}$ Since $4 \cdot 4 = 16$, $-\sqrt{16} = -4$.

3. $\sqrt{0.25}$ Since $0.5 \cdot 0.5 = 0.25$, $\sqrt{0.25} = 0.5$.

4. $\sqrt{\frac{25}{36}}$ Since $\frac{5}{6} \cdot \frac{5}{6} = \frac{25}{36}$, $\sqrt{\frac{25}{36}} = \frac{5}{6}$.

Example 5 **Solve $a^2 = \frac{4}{9}$.**

$a^2 = \frac{4}{9}$ Write the equation.

$\sqrt{a^2} = \sqrt{\frac{4}{9}}$ Take the square root of each side.

$a = \frac{2}{3}$ or $-\frac{2}{3}$ Notice that $\frac{2}{3} \cdot \frac{2}{3} = \frac{4}{9}$ and $\left(-\frac{2}{3}\right)\left(-\frac{2}{3}\right) = \frac{4}{9}$.

The equation has two solutions, $\frac{2}{3}$ and $-\frac{2}{3}$.

Exercises

Find each square root.

1. $\sqrt{4}$

2. $\sqrt{9}$

3. $-\sqrt{49}$

4. $-\sqrt{25}$

5. $\sqrt{0.01}$

6. $-\sqrt{0.64}$

7. $\sqrt{\frac{9}{16}}$

8. $-\sqrt{\frac{1}{25}}$

ALGEBRA Solve each equation.

9. $x^2 = 121$

10. $a^2 = 3{,}600$

11. $p^2 = \frac{81}{100}$

12. $t^2 = \frac{121}{196}$

Lesson 3-1

NAME ______________________ DATE __________ PERIOD _____

3-1 Practice

Square Roots

Find each square root.

1. $\sqrt{36}$

2. $-\sqrt{144}$

3. $-\sqrt{\frac{9}{16}}$

4. $\sqrt{1.96}$

5. $\pm\sqrt{2.25}$

6. $\pm\sqrt{\frac{121}{289}}$

7. $\sqrt{\frac{81}{100}}$

8. $\pm\sqrt{0.0025}$

9. $-\sqrt{0.49}$

10. $-\sqrt{3.24}$

11. $-\sqrt{\frac{25}{441}}$

12. $\pm\sqrt{361}$

ALGEBRA Solve each equation. Check your solution(s).

13. $h^2 = 121$

14. $324 = a^2$

15. $x^2 = \frac{81}{169}$

16. $0.0196 = m^2$

17. $\sqrt{y} = 6$

18. $\sqrt{z} = 8.4$

19. GARDENING Moesha has 196 pepper plants that she wants to plant in square formation. How many pepper plants should she plant in each row?

20. RESTAURANTS A new restaurant has ordered 64 tables for its outdoor patio. If the manager arranges the tables in a square formation, how many will be in each row?

GEOMETRY The formula for the perimeter of a square is $P = 4s$, where s is the length of a side. Find the perimeter of each square.

21. Area = 144 square inches

22. Area = 81 square feet

23. Area = 324 square meters

NAME ______________________________ DATE ____________ PERIOD _____

3-2 Study Guide and Intervention

Estimating Square Roots

Most numbers are not perfect squares. You can estimate square roots for these numbers.

Example 1 **Estimate $\sqrt{204}$ to the nearest whole number.**

- The first perfect square less than 204 is 196.
- The first perfect square greater than 204 is 225.

$196 < 204 < 225$ Write an inequality.

$14^2 < 204 < 15^2$ $196 = 14^2$ and $225 = 15^2$

$14 < \sqrt{204} < 15$ Take the square root of each number.

So, $\sqrt{204}$ is between 14 and 15. Since 204 is closer to 196 than 225, the best whole number estimate for $\sqrt{204}$ is 14.

Example 2 **Estimate $\sqrt{79.3}$ to the nearest whole number.**

- The first perfect square less than 79.3 is 64.
- The first perfect square greater than 79.3 is 81.

$64 < 79.3 < 81$ Write an inequality.

$8^2 < 79.3 < 9^2$ $64 = 8^2$ and $81 = 9^2$

$8 < \sqrt{79.3} < 9$ Take the square root of each number.

So, $\sqrt{79.3}$ is between 8 and 9. Since 79.3 is closer to 81 than 64, the best whole number estimate for $\sqrt{79.3}$ is 9.

Exercises

Estimate to the nearest whole number.

1. $\sqrt{8}$ **2.** $\sqrt{37}$ **3.** $\sqrt{14}$

4. $\sqrt{26}$ **5.** $\sqrt{62}$ **6.** $\sqrt{48}$

7. $\sqrt{103}$ **8.** $\sqrt{141}$ **9.** $\sqrt{14.3}$

10. $\sqrt{51.2}$ **11.** $\sqrt{82.7}$ **12.** $\sqrt{175.2}$

Lesson 3-2

NAME ______________________________ DATE ____________ PERIOD ______

3-2 Practice

Estimating Square Roots

Estimate to the nearest whole number.

1. $\sqrt{38}$ **2.** $\sqrt{53}$ **3.** $\sqrt{99}$ **4.** $\sqrt{227}$

5. $\sqrt{8.5}$ **6.** $\sqrt{35.1}$ **7.** $\sqrt{67.3}$ **8.** $\sqrt{103.6}$

9. $\sqrt{86.4}$ **10.** $\sqrt{45.2}$ **11.** $\sqrt{7\frac{2}{5}}$ **12.** $\sqrt{27\frac{3}{8}}$

Order from least to greatest.

13. 8, 10, $\sqrt{61}$, $\sqrt{73}$ **14.** $\sqrt{45}$, 9, 6, $\sqrt{63}$ **15.** $\sqrt{50}$, 7, $\sqrt{44}$, 5

ALGEBRA **Estimate the solution of each equation to the nearest integer.**

16. $d^2 = 61$ **17.** $z^2 = 85$ **18.** $r^2 = 3.7$

19. GEOMETRY The radius of a cylinder with volume V and height 10 centimeters is approximately $\sqrt{\frac{V}{30}}$. If a can that is 10 centimeters tall has a volume of 900 cubic centimeters, estimate its radius.

20. TRAVEL The formula $s = \sqrt{18d}$ can be used to find the speed s of a car in miles per hour when the car needs d feet to come to a complete stop after slamming on the brakes. If it took a car 12 feet to come to a complete stop after slamming on the brakes, estimate the speed of the car.

GEOMETRY **The formula for the area of a square is $A = s^2$, where s is the length of a side. Estimate the length of a side for each square.**

21. Area = 40 square inches

22. Area = 97 square feet

3-3 Study Guide and Intervention

Problem-Solving Investigation: Use a Venn Diagram

You may need to use a Venn diagram to solve some problems.

Understand	• Determine what information is given in the problem and what you need to find.
Plan	• Select a strategy including a possible estimate.
Solve	• Solve the problem by carrying out your plan.
Check	• Examine your answer to see if it seems reasonable.

Example

Of the 25 skiers on the ski team, 13 signed up to race in the Slalom race, and 8 signed up for the Giant Slalom race. Six skiers signed up to ski in both the Slalom and the Giant Slalom races. How many skiers did not sign up for any races?

Understand You know how many skiers signed up for each race and how many signed up for both races. You need to organize the information.

Plan You can use a Venn diagram to organize the information.

Solve Draw two overlapping circles to represent the two different races. Place a 6 in the section that is a part of both circles. Use subtraction to determine the number for each other section.

only the Slalom race: $13 - 6 = 7$
only the Giant Slalom race: $8 - 6 = 2$
neither the Slalom or the Giant Slalom race:
$25 - 7 - 2 - 6 = 10$

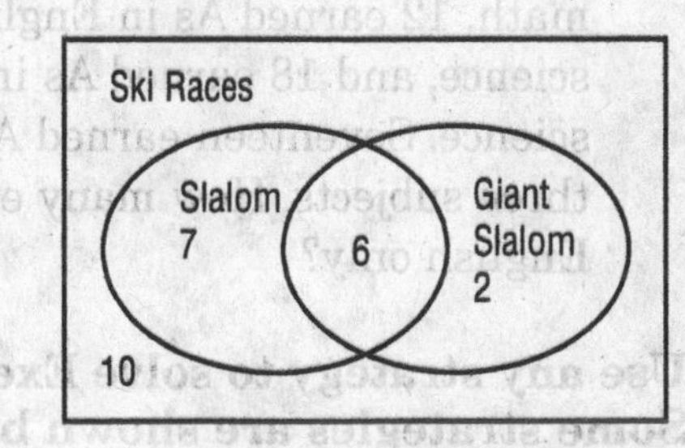

There were 10 skiers who did not sign up for either race.

Check Check each circle to see if the appropriate number of students is represented.

Exercise

Use a Venn diagram to solve the problem.

SPORTS The athletic club took a survey to find out what sports students might participate in next fall. Of the 80 students surveyed, 42 wanted to play football, 37 wanted to play soccer, and 15 wanted to play both football and soccer. How many students did not want to play either sport in the fall?

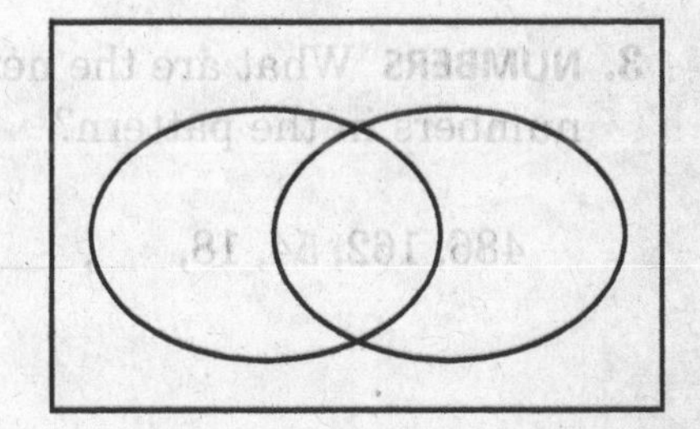

NAME ______________________ DATE ____________ PERIOD _____

3-3 Practice

Problem-Solving Investigation: Use a Venn Diagram

Mixed Problem Solving

Use a Venn diagram to solve Exercises 1 and 2.

1. **SPORTS** Of the 25 baseball players on the Baltimore Orioles 2005 roster, 17 threw right handed, 12 were over 30 years old, and 9 both threw right handed and were over 30 years old. How many players on the team neither threw right handed nor were over 30 years old?

2. **GRADES** The principal noticed that 45 students earned As in English, 49 students earned As in math, and 53 students earned As in science. Of those who earned As in exactly two of the subjects, 8 earned As in English and math, 12 earned As in English and science, and 18 earned As in math and science. Seventeen earned As in all three subjects. How many earned As in English only?

Use any strategy to solve Exercises 3–6. Some strategies are shown below.

PROBLEM-SOLVING STRATEGIES
• Look for a pattern.
• Use a Venn diagram.
• Guess and check.

3. **NUMBERS** What are the next two numbers in the pattern?

   486, 162, 54, 18, ___, ___

4. **GEOGRAPHY** Of the 50 U.S. states, 30 states border a major body of water and 14 states border a foreign country. Seven states border both a major body of water and a foreign country. How many states border on just a major body of water and how many border on just a foreign country?

5. **LANDSCAPING** Three different landscaping companies treat lawns for weeds. Company A charges $35 per treatment and requires 3 treatments to get rid of weeds. Company B charges $30 per treatment and requires 4 treatments. Company C charges $50 per treatment and requires only two treatments to eliminate weeds. If you want to use the company that charges the least, which company should you choose?

6. **RECEIVING** Marc unloaded 7,200 bottles of water from delivery trucks today. If each truck contained 50 cases and each case contained 24 bottles of water, how many trucks did he unload?

3-4 Study Guide and Intervention

The Real Number System

Numbers may be classified by identifying to which of the following sets they belong.

Whole Numbers	0, 1, 2, 3, 4, ...
Integers	..., −2, −1, 0, 1, 2, ...
Rational Numbers	numbers that can be expressed in the form $\frac{a}{b}$, where a and b are integers and $b \neq 0$
Irrational Numbers	numbers that cannot be expressed in the form $\frac{a}{b}$, where a and b are integers and $b \neq 0$

Examples **Name all sets of numbers to which each real number belongs.**

1. 5 — whole number, integer, rational number
2. 0.666... — Decimals that terminate or repeat are rational numbers, since they can be expressed as fractions. $0.666... = \frac{2}{3}$
3. $-\sqrt{25}$ — Since $-\sqrt{25} = -5$, it is an integer and a rational number.
4. $-\sqrt{11}$ — $\sqrt{11} \approx 3.31662479...$ Since the decimal does not terminate or repeat, it is an irrational number.

To compare real numbers, write each number as a decimal and then compare the decimal values.

Example 5 **Replace ● with <, >, or = to make $2\frac{1}{4}$ ● $\sqrt{5}$ a true sentence.**

Write each number as a decimal.

$2\frac{1}{4} = 2.25$

$\sqrt{5} \approx 2.236067...$

Since 2.25 is greater than 2.236067..., $2\frac{1}{4} > \sqrt{5}$.

Exercises

Name all sets of numbers to which each real number belongs.

1. 30

2. −11

3. $5\frac{4}{7}$

4. $\sqrt{21}$

5. 0

6. $-\sqrt{9}$

7. $\frac{6}{3}$

8. $-\sqrt{101}$

Replace each ● with <, >, or = to make a true sentence.

9. 2.7 ● $\sqrt{7}$

10. $\sqrt{11}$ ● $3\frac{1}{2}$

11. $4\frac{1}{6}$ ● $\sqrt{17}$

12. $3.\overline{8}$ ● $\sqrt{15}$

NAME ______________________________ DATE ____________ PERIOD _____

3-4 Practice

The Real Number System

Name all sets of numbers to which the real number belongs.

1. -9 **2.** $\sqrt{144}$ **3.** $\sqrt{35}$ **4.** $\frac{8}{11}$

5. 9.55 **6.** $5.\overline{3}$ **7.** $\frac{20}{5}$ **8.** $-\sqrt{44}$

Estimate each square root to the nearest tenth. Then graph the square root on a number line.

9. $\sqrt{7}$ **10.** $\sqrt{19}$ **11.** $-\sqrt{33}$

Replace each ● with <, >, or = to make a true sentence.

12. $\sqrt{8}$ ● 2.7 **13.** $\sqrt{15}$ ● 3.9 **14.** $5\frac{2}{5}$ ● $\sqrt{30}$

15. $2\frac{3}{10}$ ● $\sqrt{5.29}$ **16.** $\sqrt{9.8}$ ● $3.\overline{1}$ **17.** $8.\overline{2}$ ● $8\frac{2}{9}$

Order each set of numbers from least to greatest.

18. $\sqrt{10}$, $\sqrt{8}$, 2.75, $2.\overline{8}$ **19.** 5.01, $5.0\overline{1}$, $5.\overline{01}$, $\sqrt{26}$ **20.** $-\sqrt{12}$, $\sqrt{13}$, -3.5, 3.5

21. ALGEBRA The *geometric mean* of two numbers a and b is $\sqrt{ab}$. Find the geometric mean of 32 and 50.

22. ART The area of a square painting is 600 square inches. To the nearest hundredth inch, what is the perimeter of the painting?

3-5 Study Guide and Intervention

The Pythagorean Theorem

The **Pythagorean Theorem** describes the relationship between the lengths of the legs of any right triangle. In a right triangle, the square of the length of the hypotenuse is equal to the sum of the squares of the lengths of the legs. You can use the Pythagorean Theorem to find the length of a side of a right triangle if the lengths of the other two legs are known.

Examples **Find the missing measure for each right triangle. Round to the nearest tenth if necessary.**

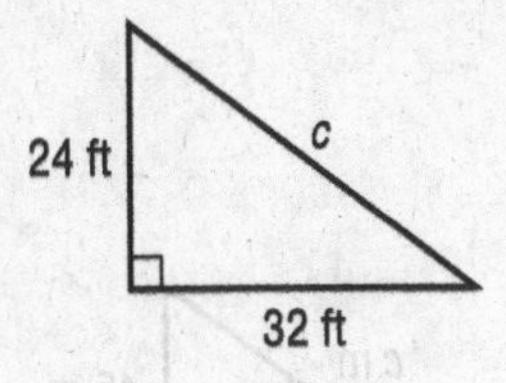

$c^2 = a^2 + b^2$
$c^2 = 24^2 + 32^2$
$c^2 = 576 + 1{,}024$
$c^2 = 1{,}600$
$c = \pm\sqrt{1{,}600}$
$c = 40 \text{ or } -40$

Length must be positive, so the length of the hypotenuse is 40 feet.

2

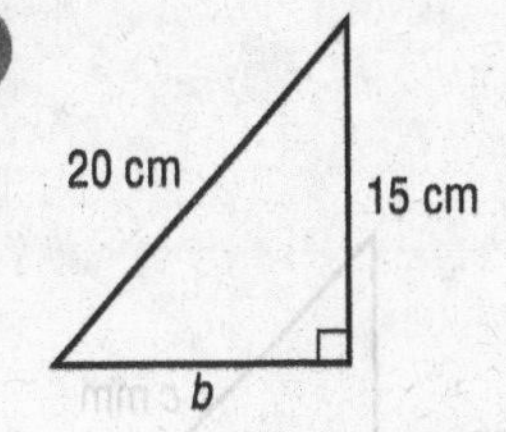

$c^2 = a^2 + b^2$
$20^2 = 15^2 + b^2$
$400 = 225 + b^2$
$400 - 225 = 225 + b^2 - 225$
$175 = b^2$
$\sqrt{175} = \sqrt{b^2}$
$13.2 \approx b$

The length of the other leg is about 13.2 centimeters.

Exercises

Write an equation you could use to find the length of the missing side of each right triangle. Then find the missing length. Round to the nearest tenth if necessary.

1.

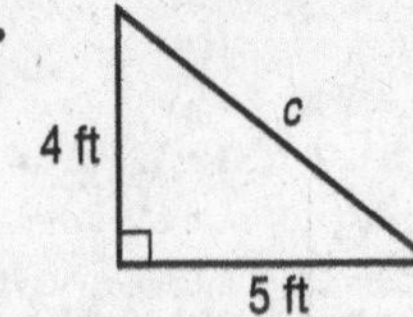

2.

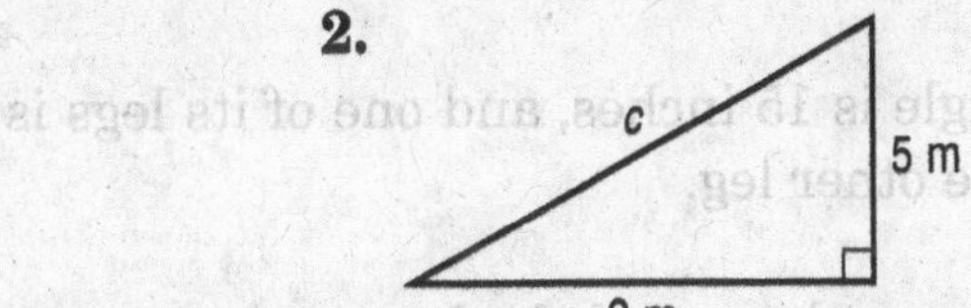

3.

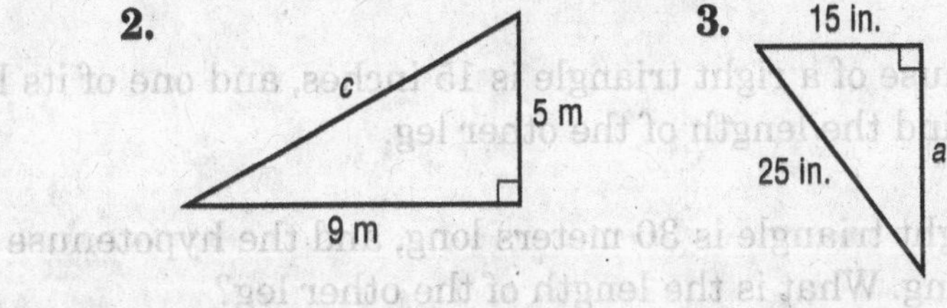

4. $a = 7$ km, $b = 12$ km

5. $a = 10$ yd, $c = 25$ yd

6. $b = 14$ ft, $c = 20$ ft

NAME ______________________________ DATE ______________ PERIOD _____

3-5 Practice

The Pythagorean Theorem

Write an equation you could use to find the length of the missing side of each right triangle. Then find the missing length. Round to the nearest tenth if necessary.

1.

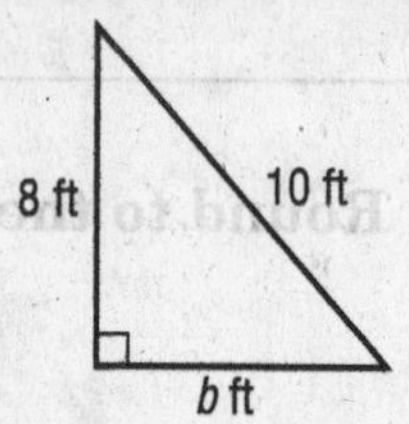

2.

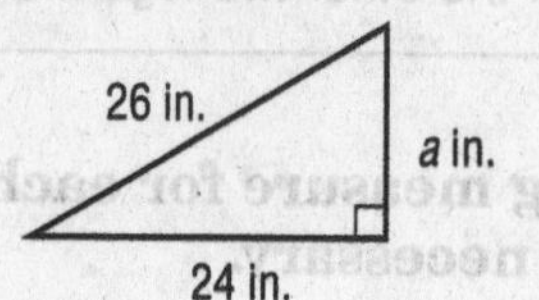

3.

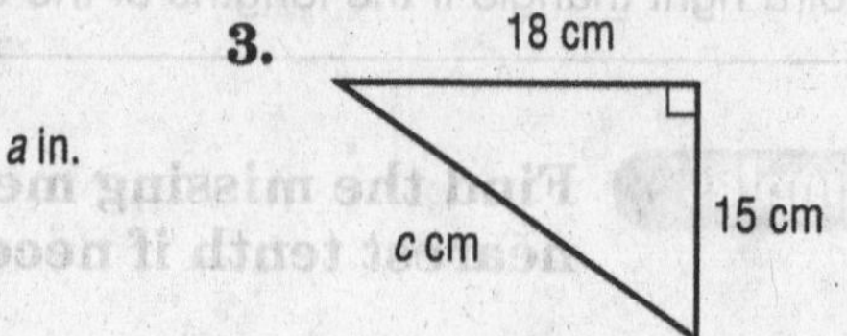

4.

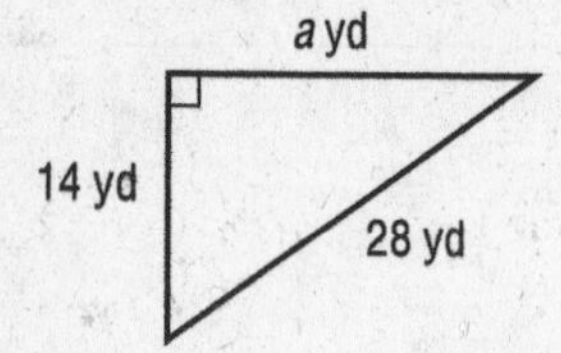

5.

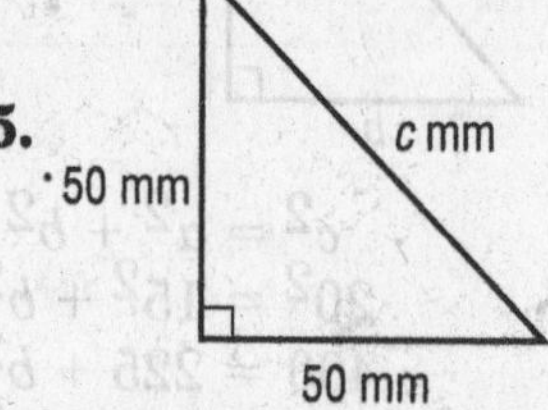

6.

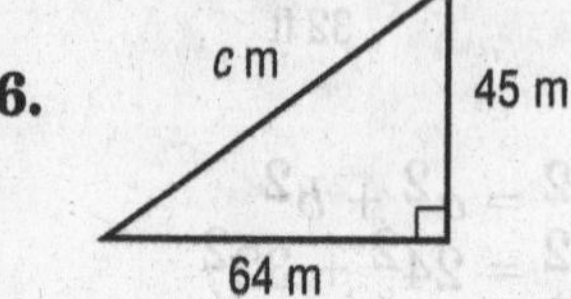

7. a, 65 cm; c, 95 cm

8. a, 16 yd; b, 22 yd

Determine whether each triangle with sides of given lengths is a right triangle.

9. 18 ft, 23 ft, 29 ft

10. 7 yd, 24 yd, 25 yd

11. The hypotenuse of a right triangle is 15 inches, and one of its legs is 11 inches. Find the length of the other leg.

12. A leg of a right triangle is 30 meters long, and the hypotenuse is 35 meters long. What is the length of the other leg?

13. **TELEVISIONS** The diagonal of a 27-inch television measures 27 inches. If the width of a 27-inch is 22 inches, calculate its height to the nearest inch.

3-6 Study Guide and Intervention

Using the Pythagorean Theorem

You can use the Pythagorean Theorem to help you solve problems.

Example 1 **A professional ice hockey rink is 200 feet long and 85 feet wide. What is the length of the diagonal of the rink?**

$c^2 = a^2 + b^2$	The Pythagorean Theorem
$c^2 = 200^2 + 85^2$	Replace *a* with 200 and *b* with 85.
$c^2 = 40{,}000 + 7{,}225$	Evaluate 200^2 and 85^2.
$c^2 = 47{,}225$	Simplify.
$\sqrt{c^2} = \sqrt{47{,}225}$	Take the square root of each side.
$c \approx 217.3$	Simplify.

The length of the diagonal of an ice hockey rink is about 217.3 feet.

Exercises

Write an equation that can be used to answer the question. Then solve. Round to the nearest tenth if necessary.

1. What is the length of the diagonal?

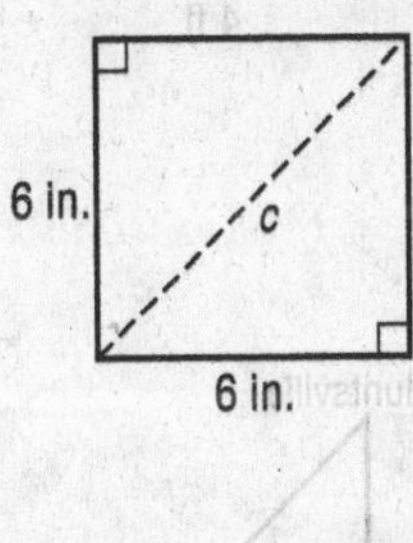

2. How long is the kite string?

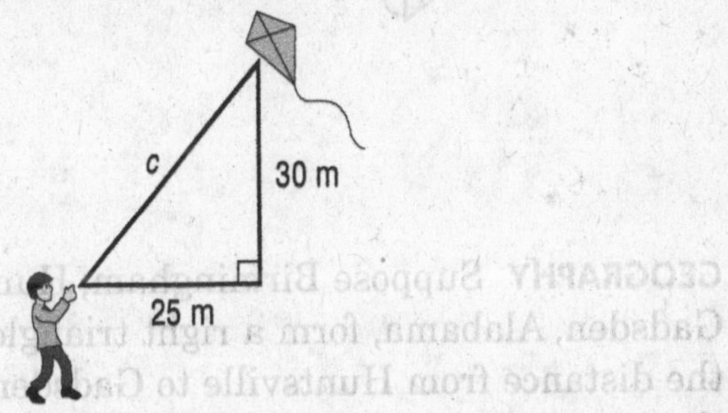

3. How high is the ramp?

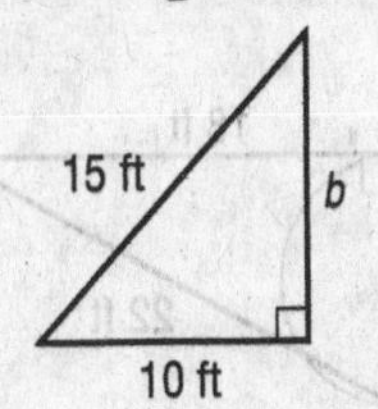

4. How tall is the tree?

NAME ______________________ DATE __________ PERIOD _____

3-6 Practice

Using the Pythagorean Theorem

Write an equation that can be used to answer the question. Then solve. Round to the nearest tenth if necessary.

1. How far is the ship from the lighthouse?

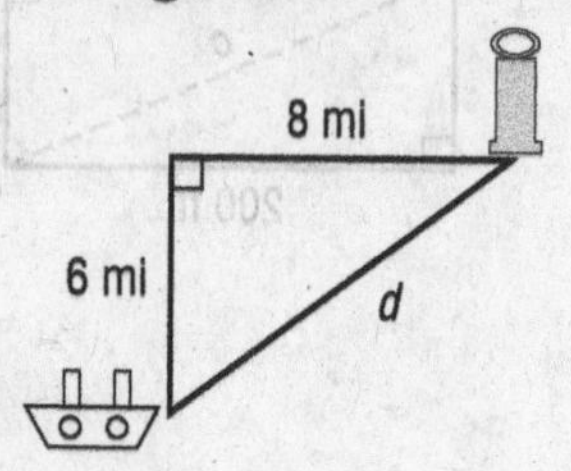

2. How long is the wire supporting the sign?

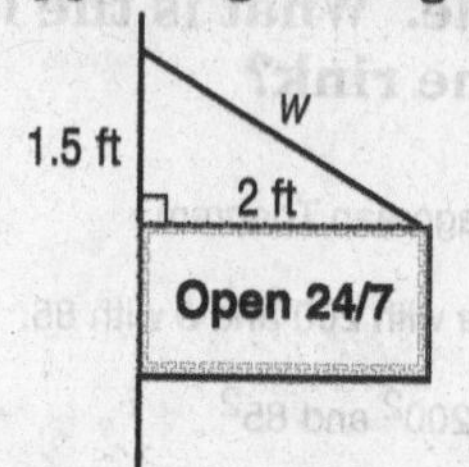

3. How far above the water is the person parasailing?

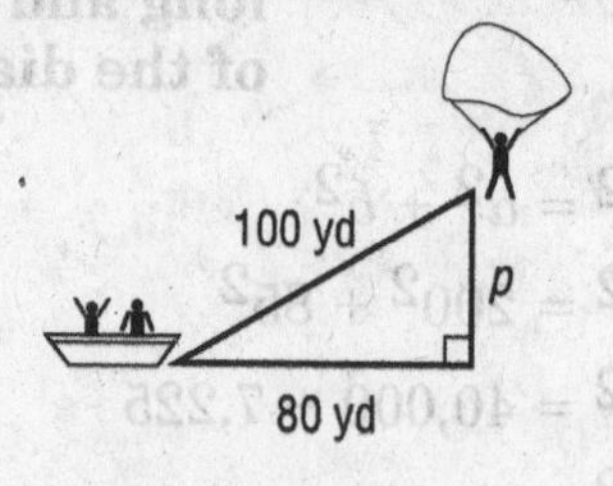

4. How wide is the pond?

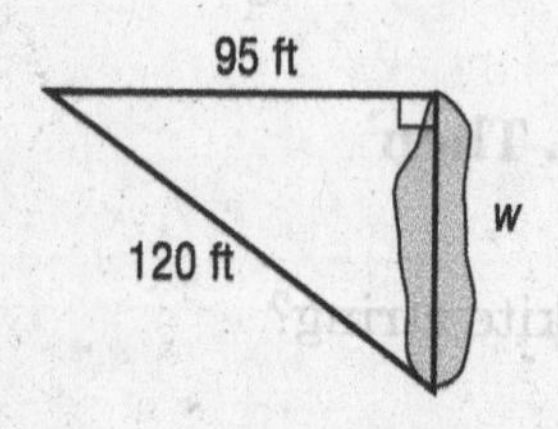

5. How high is the ramp?

6. How high is the end of the ladder against the building?

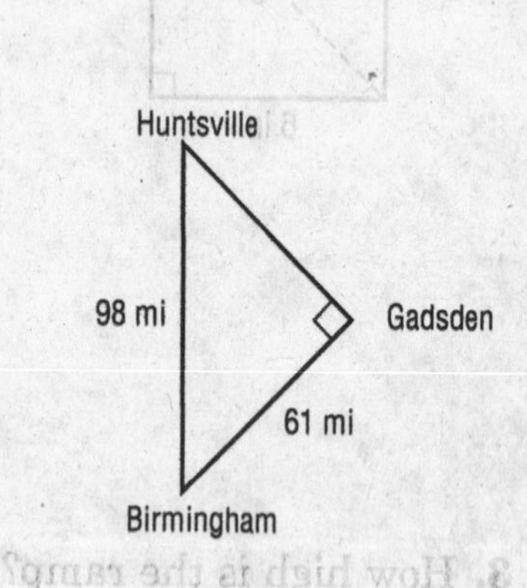

7. GEOGRAPHY Suppose Birmingham, Huntsville, and Gadsden, Alabama, form a right triangle. What is the distance from Huntsville to Gadsden? Round to the nearest tenth if necessary.

Huntsville

98 mi

Gadsden

61 mi

Birmingham

8. GEOMETRY Find the diameter d of the circle in the figure at the right. Round to the nearest tenth if necessary.

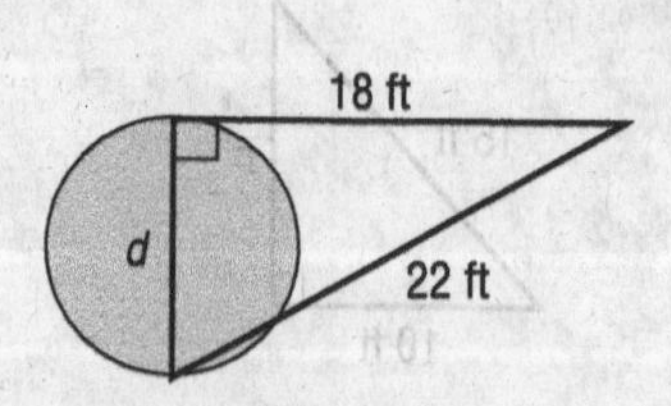

NAME ______________________ DATE ____________ PERIOD ______

3-7 Study Guide and Intervention

Distance on the Coordinate Plane

You can use the Pythagorean Theorem to find the distance between two points on the coordinate plane.

Example **Find the distance between points (2, −3) and (5, 4).**

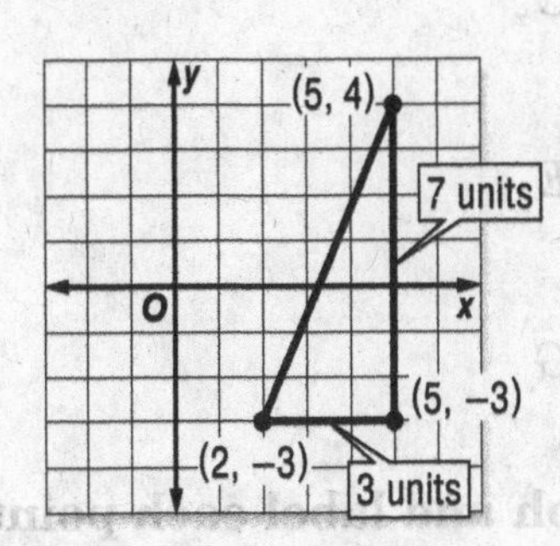

Graph the points and connect them with a line segment. Draw a horizontal line through (2, −3) and a vertical line through (5, 4). The lines intersect at (5, −3).

Count units to find the length of each leg of the triangle. The lengths are 3 units and 7 units. Then use the Pythagorean Theorem to find the hypotenuse.

$c^2 = a^2 + b^2$	The Pythagorean Theorem
$c^2 = 3^2 + 7^2$	Replace a with 3 and b with 7.
$c^2 = 9 + 49$	Evaluate 3^2 and 7^2.
$c^2 = 58$	Simplify.
$\sqrt{c^2} = \sqrt{58}$	Take the square root of each side.
$c \approx 7.6$	Simplify.

The distance between the points is about 7.6 units.

Exercises

Find the distance between each pair of points whose coordinates are given. Round to the nearest tenth if necessary.

1.

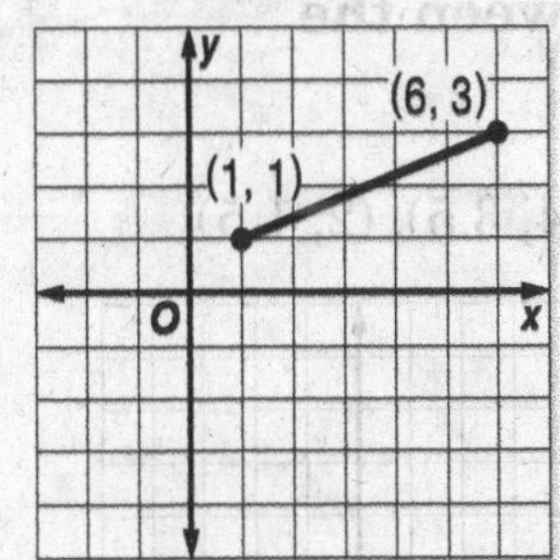

2.

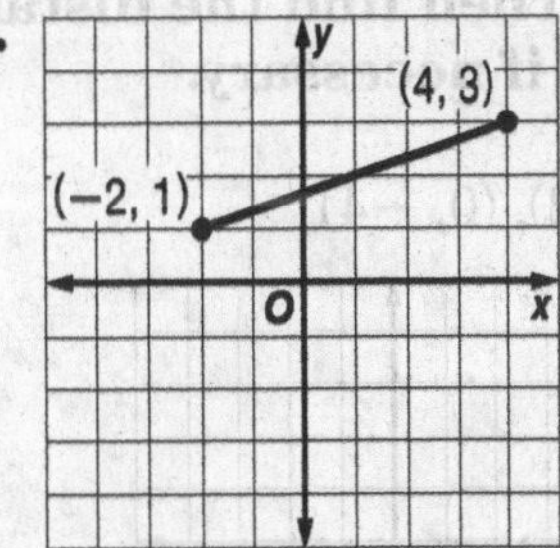

3.

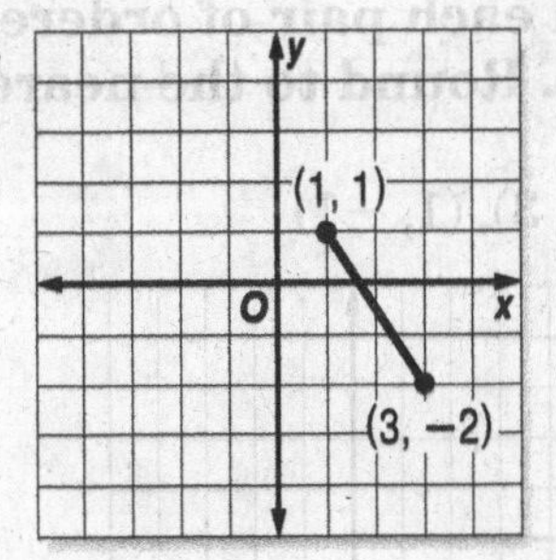

Graph each pair of ordered pairs. Then find the distance between the points. Round to the nearest tenth if necessary.

4. (4, 5), (0, 2)

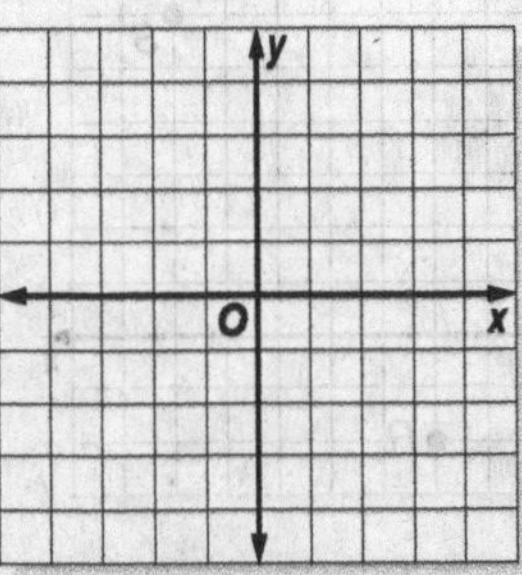

5. (0, −4), (−3, 0)

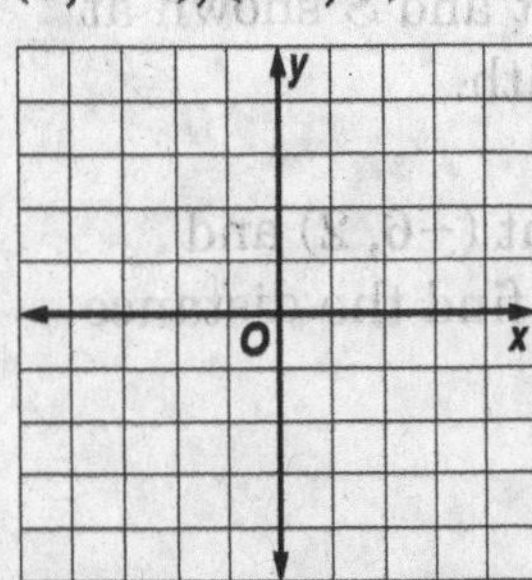

6. (−1, 1), (−4, 4)

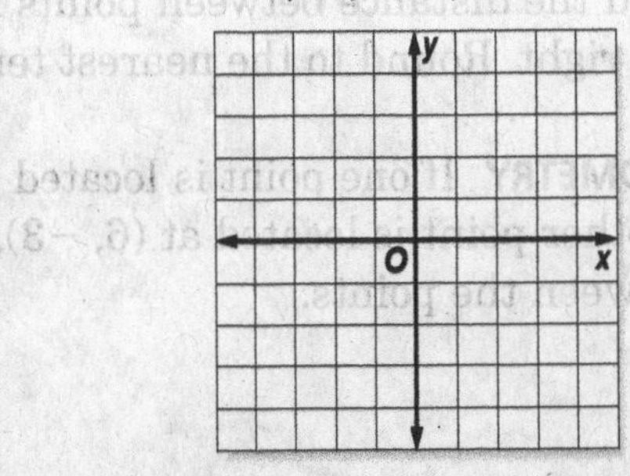

Lesson 3-7

3-7 Practice

Distance on the Coordinate Plane

Name the ordered pair for each point.

1. *A*
2. *B*
3. *C*
4. *D*
5. *E*
6. *F*
7. *G*
8. *H*

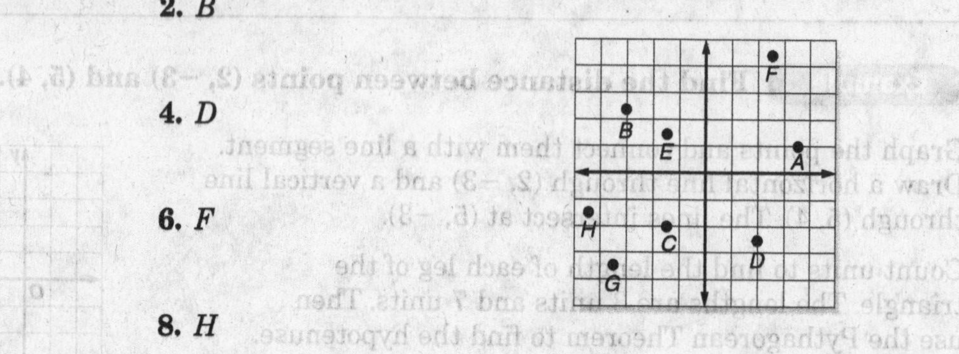

Graph and label each point.

9. $J\left(2\frac{1}{4}, \frac{1}{2}\right)$
10. $K\left(3, -1\frac{2}{3}\right)$
11. $M\left(-3\frac{3}{4}, 4\frac{1}{4}\right)$
12. $N\left(-3\frac{2}{5}, -2\frac{3}{5}\right)$
13. $P(-2.1, 1.8)$
14. $Q(1.75, -3.5)$

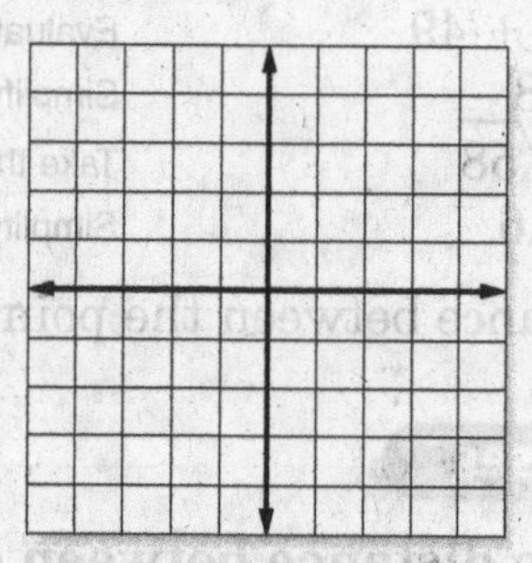

Graph each pair of ordered pairs. Then find the distance between the points. Round to the nearest tenth if necessary.

15. (4, 3), (1, −1)

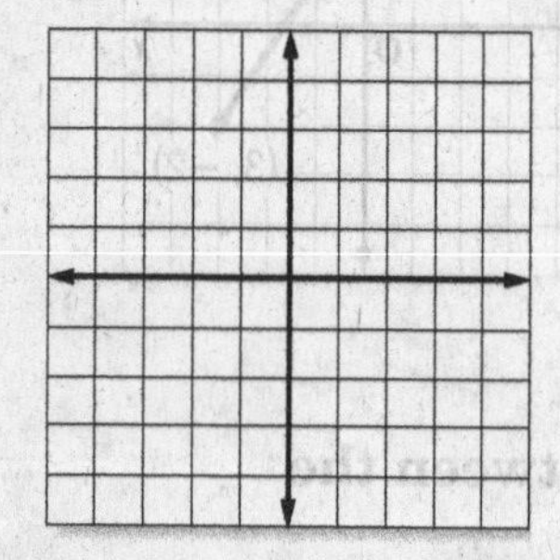

16. (3, 2), (0, −4)

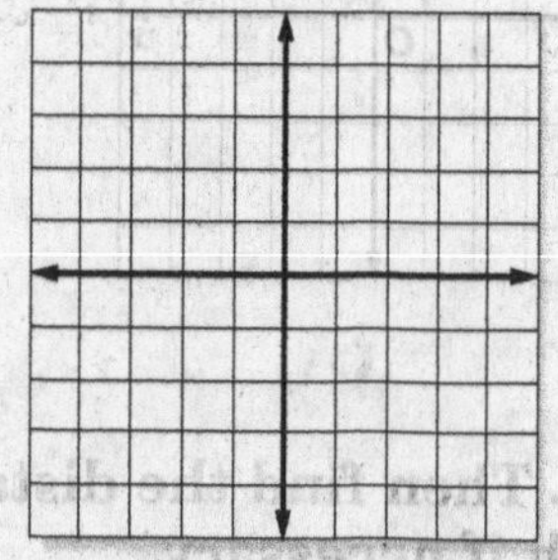

17. (−4, 3.5), (2, 1.5)

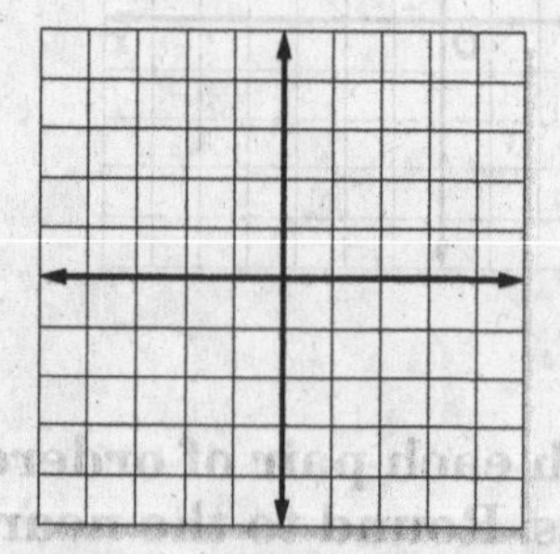

18. Find the distance between points *R* and *S* shown at the right. Round to the nearest tenth.

19. **GEOMETRY** If one point is located at (−6, 2) and another point is located at (6, −3), find the distance between the points.

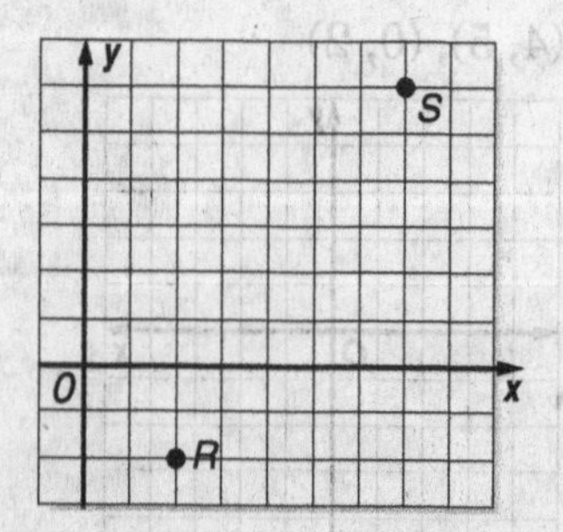

NAME ______________________ DATE __________ PERIOD _____

4-1 Study Guide and Intervention

Ratios and Rates

A **ratio** is a comparison of two numbers by quantities. Since a ratio can be written as a fraction, it can be simplified.

Example 1 **Express *35 wins to 42 losses* in simplest form.**

$\frac{35}{42} = \frac{5}{6}$ Divide the numerator and denominator by the greatest common factor, 7.

The ratio in simplest form is $\frac{5}{6}$ or 5:6.

Example 2 **Express *1 foot to 3 inches* in simplest form.**

To simplify a ratio involving measurements, both quantities must have the same unit of measure.

$\frac{\text{1 foot}}{\text{3 inches}} = \frac{\text{12 inches}}{\text{3 inches}}$ Convert 1 foot to 12 inches.

$= \frac{\text{4 inches}}{\text{1 inch}}$ Divide the numerator and denominator by 3.

The ratio in simplest form is $\frac{4}{1}$ or 4:1.

A **rate** is a ratio that compares two quanitities with different types of units. A unit rate is a rate with a denominator of 1.

Example 3 **Express *309 miles in 6 hours* as a unit rate.**

$\frac{\text{309 miles}}{\text{6 hours}} = \frac{\text{51.5 miles}}{\text{1 hour}}$ Divide the numerator and denominator by 6 to get a denominator of 1.

The unit rate is 51.5 miles per hour.

Exercises

Express each ratio in simplest form.

1. 3 out of 9 students

2. 8 passengers:2 cars

3. 5 out of 10 dentists

4. 35 boys:60 girls

5. 18 red apples to 42 green apples

6. 50 millimeters to 1 meter

Express each rate as a unit rate.

7. 12 waves in 2 hours

8. 200 miles in 4 hours

9. 21 gallons in 2.4 minutes

10. $12 for 4.8 pounds

Lesson 4-1

NAME ______________________ DATE __________ PERIOD _____

4-1 Practice

Ratios and Rates

Express each ratio in simplest form.

1. 32 out of 200 adults like opera

2. 20 picked out of 65 who tried out

3. 48 robins to 21 blackbirds seen

4. 10 rock musicians to 22 classical musicians in the concert

5. 2 feet long to 64 inches wide

6. 45 millimeters out of 10 centimeters

7. 10 ounces sugar for 1 pound apples

8. 2 quarts out of 4 gallons leaked out

Express each rate as a unit rate.

9. 110 inches of snow in 8 days

10. 38 feet in 25 seconds

11. 594 cars crossing the bridge in 3 hours

12. 366 miles on 12 gallons

13. SHOPPING An 8-ounce box of Crispy Crackers costs $1.59 and a 2-pound box costs $6.79. Which box is the better buy? Explain your reasoning.

14. ANIMALS Which animal listed in the table consumes the least amount of food compared to its body weight? Explain your reasoning.

Animal	Body Weight (lb)	Amount of Food per Day (lb)
African Elephant	12,000	500
Blue Whale	286,000	8,000
Koala	22	2
Komodo Dragon	300	240

Source: *Scholastic Book of World Records*

4-2 Study Guide and Intervention

Proportional and Nonproportional Relationships

Two related quantities are **proportional** if they have a constant ratio between them. If two related quantities do not have a constant ratio, then they are **nonproportional.**

Example 1 **The cost of one CD at a record store is $12. Create a table to show the total cost for different numbers of CDs. Is the total cost proportional to the number of CDs purchased?**

Number of CDs	1	2	3	4
Total Cost	$12	$24	$36	$48

$\frac{\text{Total Cost}}{\text{Number of CDs}} = \frac{12}{1} = \frac{24}{2} = \frac{36}{3} = \frac{48}{4} = \$12$ per CD   Divide the total cost for each by the number of CDs to find a ratio. Compare the ratios.

Since the ratios are the same, the total cost is proportional to the number of CDs purchased.

Example 2 **The cost to rent a lane at a bowling alley is $9 per hour plus $4 for shoe rental. Create a table to show the total cost for the number of hours a bowling lane is rented if one person rents shoes. Is the total cost proportional to the number of hours rented?**

Number of Hours	1	2	3	4
Total Cost	$13	$22	$31	$40

$\frac{\text{Total Cost}}{\text{Number of Hours}} \rightarrow \frac{13}{1}$ or 13 $\frac{22}{2}$ or 11 $\frac{31}{3}$ or 10.34 $\frac{40}{4}$ or 10   Divide each cost by the number of hours.

Since the ratios are not the same, the total cost is nonproportional to the number of hours rented with shoes.

Exercises

Use a table of values to explain your reasoning.

1. **PICTURES** A photo developer charges $0.25 per photo developed. Is the total cost proportional to the number of photos developed?

2. **SOCCER** A soccer club has 15 players for every team, with the exception of two teams that have 16 players each. Is the number of players proportional to the number of teams?

NAME ______________________________ DATE ____________ PERIOD _____

4-2 Practice

Proportional and Nonproportional Relationships

For Exercises 1–3, use a table of values to explain your reasoning.

1. **ANIMALS** The world's fastest fish, a sailfish, swims at a rate of 69 miles per hour. Is the distance a sailfish swims proportional to the number of hours it swims?

FOSSILS For Exercises 2 and 3, use the following information.

In July, a paleontologist found 368 fossils at a dig. In August, she found about 14 fossils per day.

2. Is the number of fossils the paleontologist found in August proportional to the number of days she spent looking for fossils that month?

3. Is the total number of fossils found during July and August proportional to the number of days the paleontologist spent looking for fossils in August?

4-3 Study Guide and Intervention

Rate of Change

To find the rate of change between two data points, divide the difference of the *y*-coordinates by the difference of the *x*-coordinates. The rate of change between (x_1, y_1) and (x_2, y_2) is $\frac{y_2 - y_1}{x_2 - x_1}$.

Example INCOME **The graph shows Mr. Jackson's annual income between 1998 and 2006. Find the rate of change in Mr. Jackson's income between 1998 and 2001.**

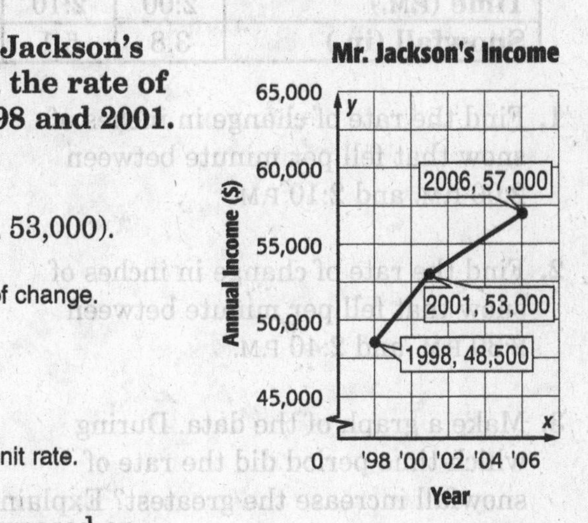

Use the formula for the rate of change.
Let $(x_1, y_1) = (1998, 48{,}500)$ and $(x_2, y_2) = (2001, 53{,}000)$.

$\frac{y_2 - y_1}{x_2 - x_1} = \frac{53{,}000 - 48{,}500}{2001 - 1998}$ Write the formula for rate of change.

$= \frac{4{,}500}{3}$ Simplify.

$= \frac{1{,}500}{1}$ Express this rate as a unit rate.

Between 1998 and 2001, Mr. Jackson's income increased an average of $1,500 per year.

Exercises

SURF **For Exercises 1–3, use the graph that shows the average daily wave height as measured by an ocean buoy over a nine-day period.**

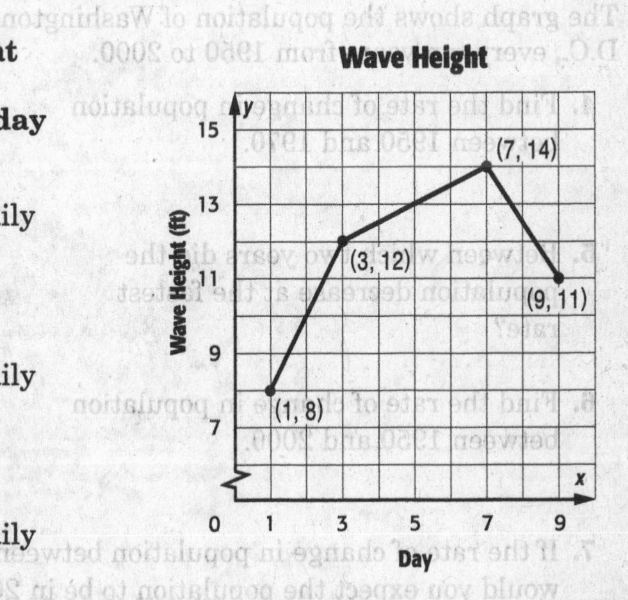

1. Find the rate of change in the average daily wave height between day 1 and day 3.

2. Find the rate of change in the average daily wave height between day 3 and day 7.

3. Find the rate of change in the average daily wave height between day 7 and day 9.

4-3 Practice

Rate of Change

SNOWFALL For Exercises 1–3, use the following information.
The amount of snow that fell during five time periods is shown in the table.

Time (P.M.)	2:00	2:10	2:20	2:30	2:40
Snowfall (in.)	3.8	5.1	5.5	7.8	8.3

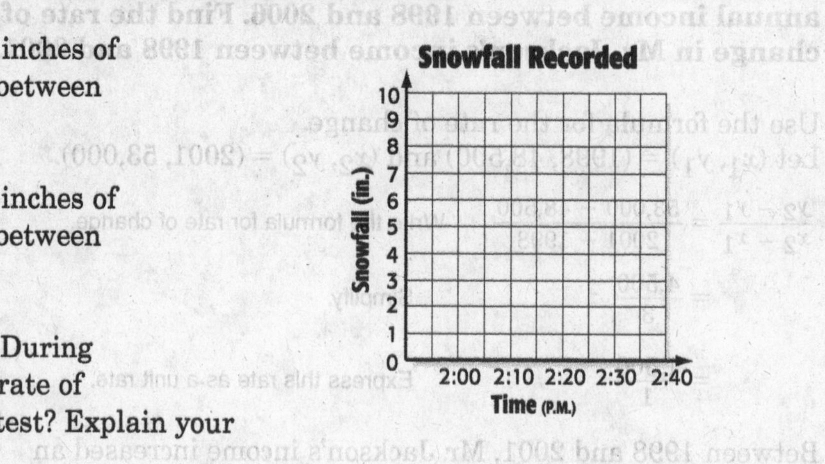

1. Find the rate of change in inches of snow that fell per minute between 2:00 P.M. and 2:10 P.M.

2. Find the rate of change in inches of snow that fell per minute between 2:30 P.M. and 2:40 P.M.

3. Make a graph of the data. During which time period did the rate of snowfall increase the greatest? Explain your reasoning.

POPULATION For Exercises 4–7, use the the information below and at the right.
The graph shows the population of Washington, D.C., every ten years from 1950 to 2000.

4. Find the rate of change in population between 1950 and 1970.

5. Between which two years did the population decrease at the fastest rate?

6. Find the rate of change in population between 1950 and 2000.

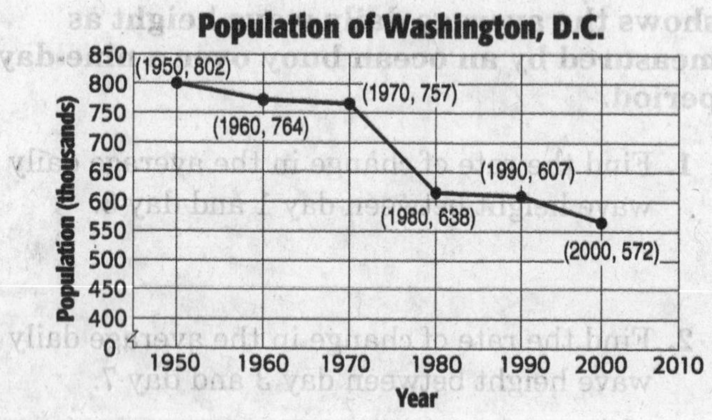

Source: U.S. Census Bureau

7. If the rate of change in population between 1950 and 2000 were to continue, what would you expect the population to be in 2010? Explain your reasoning.

4-4 Study Guide and Intervention

Constant Rate of Change

The **slope** of a line is the ratio of the rise, or vertical change, to the run, or horizontal change.

Example 1 **The table shows the relationship between feet and seconds. Is the relationship between feet and seconds linear? If so, find the constant rate of change. If not, explain your reasoning.**

$$\text{rate of change} = \frac{\text{feet}}{\text{seconds}} \begin{matrix} \leftarrow \text{change in } y \\ \leftarrow \text{change in } x \end{matrix}$$

$$= \frac{3}{-4} \text{ or } -\frac{3}{4}$$

The rate of change is $-\frac{3}{4}$ feet per second.

	+3	+3	+3	
feet	−2	1	4	−7
seconds	5	1	−3	−7
	−4	−4	−4	

Example 2 **Find the constant rate of change for the number of feet per second. Interpret its meaning.**

Choose two points on the line. The vertical change from point A to point B is 4 units while the horizontal change is 2 units.

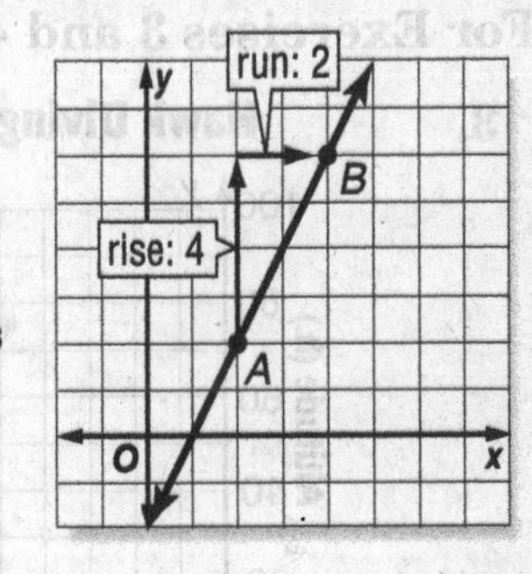

$\text{rate of change} = \frac{\text{feet}}{\text{seconds}}$	Definition of rate of change
$= \frac{4}{2}$	Difference in feet between two points divided by the difference in seconds for those two points
$= 2$	Simplify.

The rate of change is 2 feet per second.

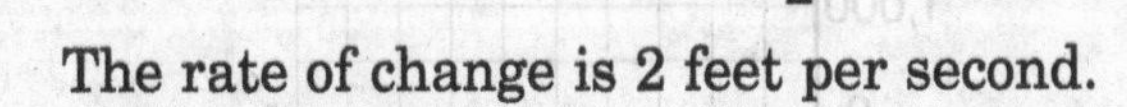

Exercises

Find the rate of change for each line.

1\.

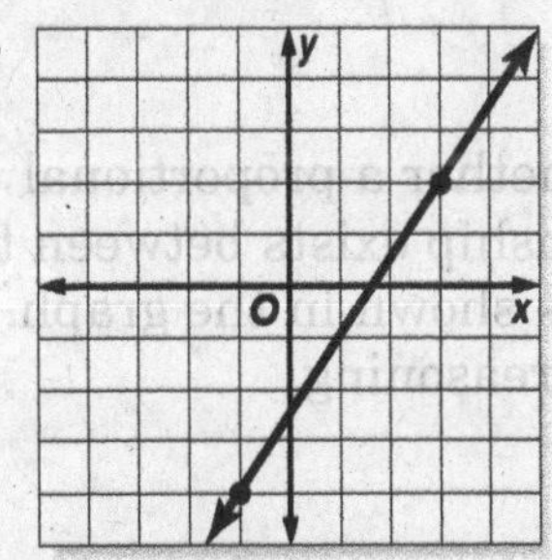

2\.

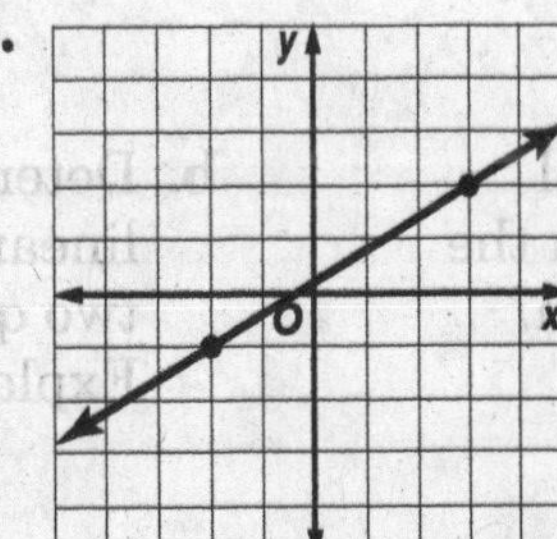

3\.

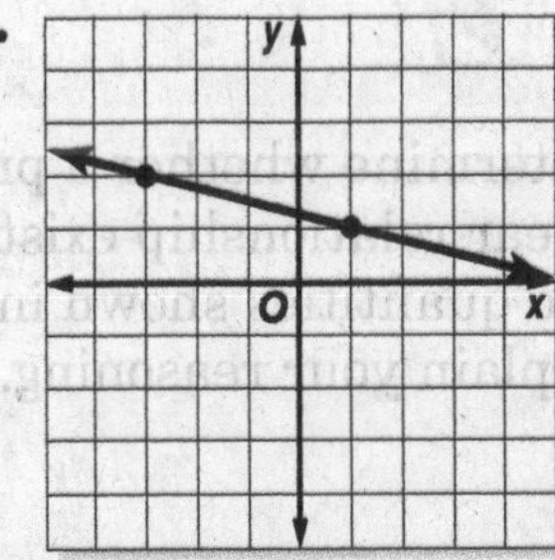

The points given in each table lie on a line. Find the rate of change for the line.

4\.

x	3	5	7	9
y	−1	2	5	8

5\.

x	−5	0	5	10
y	4	3	2	1

4-4 Practice

Constant Rate of Change

Determine whether the relationship between the two quantities described in each table is linear. If so, find the constant rate of change. If not, explain your reasoning.

1. Fabric Needed for Costumes

Number of Costumes	2	4	6	8
Fabric (yd)	7	14	21	28

2. Distance Traveled on Bike Trip

Day	1	2	3	4
Distance (mi)	21.8	43.6	68.8	90.6

For Exercises 3 and 4, refer to the graphs below.

3.

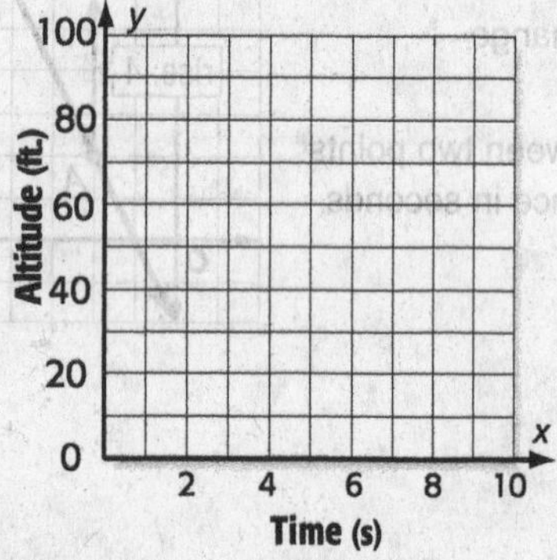

a. Find the constant rate of change and interpret its meaning.

b. Determine whether a proportional linear relationship exists between the two quantities shown in the graph. Explain your reasoning.

4.

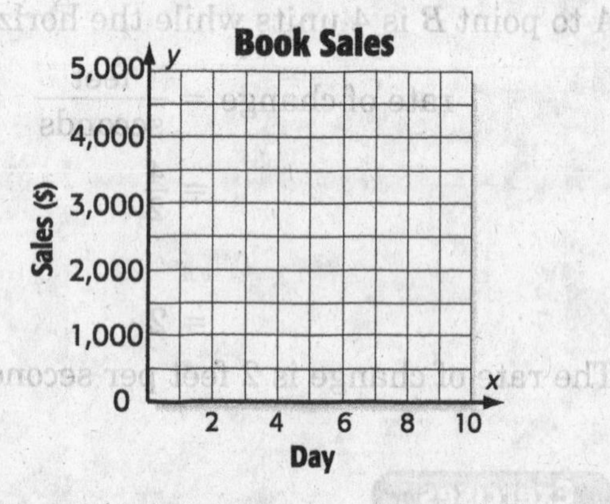

a. Find the constant rate of change and interpret its meaning.

b. Determine whether a proportional linear relationship exists between the two quantities shown in the graph. Explain your reasoning.

NAME ______________________ DATE ____________ PERIOD ______

4-5 Study Guide and Intervention

Solving Proportions

A **proportion** is an equation that states that two ratios are equivalent. To determine whether a pair of ratios forms a proportion, use cross products. You can also use cross products to solve proportions.

Example 1 **Determine whether the pair of ratios $\frac{20}{24}$ and $\frac{12}{18}$ forms a proportion.**

Find the cross products.

$\frac{20}{24} \stackrel{?}{=} \frac{12}{18}$ → $24 \cdot 12 = 288$
→ $20 \cdot 18 = 360$

Since the cross products are not equal, the ratios do not form a proportion.

Example 2 **Solve $\frac{12}{30} = \frac{k}{70}$.**

$\frac{12}{30} = \frac{k}{70}$	**Write the equation.**
$12 \cdot 70 = 30 \cdot k$	**Find the cross products.**
$840 = 30k$	**Multiply.**
$\frac{840}{30} = \frac{30k}{30}$	**Divide each side by 30.**
$28 = k$	**Simplify.** The solution is 28.

Exercises

Determine whether each pair of ratios forms a proportion.

1. $\frac{17}{10}, \frac{12}{5}$
2. $\frac{6}{9}, \frac{12}{18}$
3. $\frac{8}{12}, \frac{10}{15}$
4. $\frac{7}{15}, \frac{13}{32}$
5. $\frac{7}{9}, \frac{49}{63}$
6. $\frac{8}{24}, \frac{12}{28}$
7. $\frac{4}{7}, \frac{12}{71}$
8. $\frac{20}{35}, \frac{30}{45}$
9. $\frac{18}{24}, \frac{3}{4}$

Solve each proportion.

10. $\frac{x}{5} = \frac{15}{25}$
11. $\frac{3}{4} = \frac{12}{c}$
12. $\frac{6}{9} = \frac{10}{r}$
13. $\frac{16}{24} = \frac{z}{15}$
14. $\frac{5}{8} = \frac{s}{12}$
15. $\frac{14}{t} = \frac{10}{11}$
16. $\frac{w}{6} = \frac{2.8}{7}$
17. $\frac{5}{y} = \frac{7}{16.8}$
18. $\frac{x}{18} = \frac{7}{36}$

4-5 Practice

Solving Proportions

Solve each proportion.

1. $\frac{b}{5} = \frac{8}{16}$

2. $\frac{18}{x} = \frac{6}{10}$

3. $\frac{t}{5} = \frac{12}{80}$

4. $\frac{11}{10} = \frac{n}{14}$

5. $\frac{2.5}{35} = \frac{2}{d}$

6. $\frac{3.5}{18} = \frac{z}{36}$

7. $\frac{0.45}{4.2} = \frac{p}{14}$

8. $\frac{2.4}{6} = \frac{2.8}{s}$

9. $\frac{3.6}{k} = \frac{0.2}{0.5}$

10. **CLASSES** For every girl taking classes at the martial arts school, there are 3 boys who are taking classes at the school. If there are 236 students taking classes, write and solve a proportion to predict the number of boys taking classes at the school.

11. **BICYCLES** An assembly line worker at Rob's Bicycle factory adds a seat to a bicycle at a rate of 2 seats in 11 minutes. Write an equation relating the number of seats s to the number of minutes m. At this rate, how long will it take to add 16 seats? 19 seats?

12. **PAINTING** Lisa is painting a fence that is 26 feet long and 7 feet tall. A gallon of paint will cover 350 square feet. Write and solve a proportion to determine how many gallons of paint Lisa will need.

NAME ______________________ DATE ____________ PERIOD ______

4-6 Study Guide and Intervention

Problem-Solving Investigation: Draw a Diagram

Example **It takes a worker 4 minutes to stack 2 rows of 8 boxes in a warehouse. How long will it take to stack 8 rows of 8 boxes? Use the draw a diagram strategy to solve the problem.**

Understand After 4 minutes, a worker has stacked a 2 rows of 8 boxes. At this rate, how long would it take to stack 8 rows of boxes?

Plan Draw a diagram showing the level of boxes after 4 minutes.

Solve 2 rows of 8 boxes = 4 minutes
8 rows = 4 × 2 rows, so multiply the time by 4.
4 × 4 minutes = 16 minutes

Check

8 boxes × 2 rows of boxes = 16 boxes	Multiply to find the total number of boxes in the stack.
4 minutes ÷ 16 boxes = 0.25 min. per box	Divide the number of minutes by the number of boxes.
8 boxes × 8 rows of boxes = 64 boxes	Multiply to find the number of boxes in the new stack.
64 boxes × 0.25 min. = 16 minutes	Multiply the number of boxes by the time per box.

It will take 16 minutes to stack an 8 × 8 wall of boxes.

Exercises

For Exercises 1–4, use the draw a diagram strategy to solve the problem.

1. **GAS** A car's gas tank holds 16 gallons. After filling it for 20 seconds, the tank contains 2.5 gallons. How many more seconds will it take to fill the tank?

2. **TILING** It takes 96 tiles to fill a 2-foot by 3-foot rectangle. How many tiles would it take to fill a 4-foot by 6-foot rectangle?

3. **BEVERAGES** Four juice cartons can fill 36 glasses of juice equally. How many juice cartons are needed to fill 126 glasses equally?

4. **PACKAGING** It takes 5 large shipping boxes to hold 120 boxes of an action figure. How many action figures would 8 large shipping boxes hold?

NAME ______________________ DATE ____________ PERIOD _____

4-6 Practice

Problem-Solving Investigation: Draw a Diagram

Mixed Problem Solving

Use the draw a diagram strategy to solve Exercises 1 and 2.

1. **SWIMMING** Jon is separating the width of the swimming pool into equal-sized lanes with rope. It took him 30 minutes to create 6 equal-sized lanes. How long would it take him to create 4 equal-sized lanes in a similar swimming pool?

2. **TRAVEL** Two planes are flying from San Francisco to Chicago, a distance of 1,800 miles. They leave San Francisco at the same time. After 30 minutes, one plane has traveled 25 more miles than the other plane. How much longer will it take the slower plane to get to Chicago than the faster plane if the faster plane is traveling at 500 miles per hour?

Use any strategy to solve Exercises 3–6. Some strategies are shown below.

PROBLEM-SOLVING STRATEGIES
• Work backward. • Look for a pattern. • Use a Venn diagram. • Draw a diagram.

3. **TALENT SHOW** In a solo singing and piano playing show, 18 people sang and 14 played piano. Six people both sang and played piano. How many people were in the singing and piano playing show?

4. **LETTERS** Suppose you have three strips of paper as shown. How many capital letters of the alphabet could you form using one or more of these three strips for each letter? List them according to the number of strips.

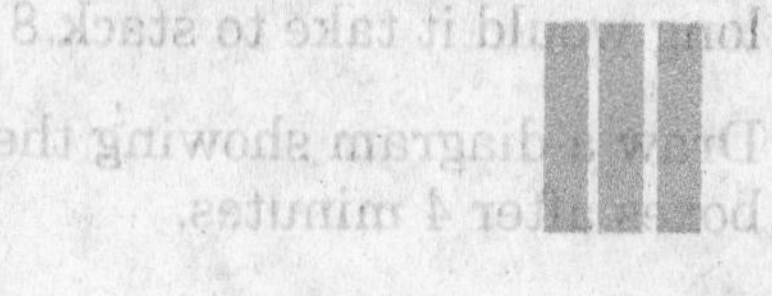

5. **CLOTHING** A store has 255 wool ponchos to sell. There are 112 adult-sized ponchos that sell for $45 each. The rest are kid-sized and sell for $32 each. If the store sells all the ponchos, how much money will the store receive?

6. **DINOSAURS** Brad made a model of a *Stegosaurus*. If you multiply the model's length by 8 and subtract 4, you will find the length of an average *Stegosaurus*. If the actual *Stegosaurus* is 30 ft long, how long is Brad's model?

NAME ______________________ DATE ____________ PERIOD _____

4-7 Study Guide and Intervention

Similar Polygons

Two polygons are **similar** if their corresponding angles are congruent and their corresponding side measures are proportional.

Example 1 **Determine whether $\triangle ABC$ is similar to $\triangle DEF$. Explain your reasoning.**

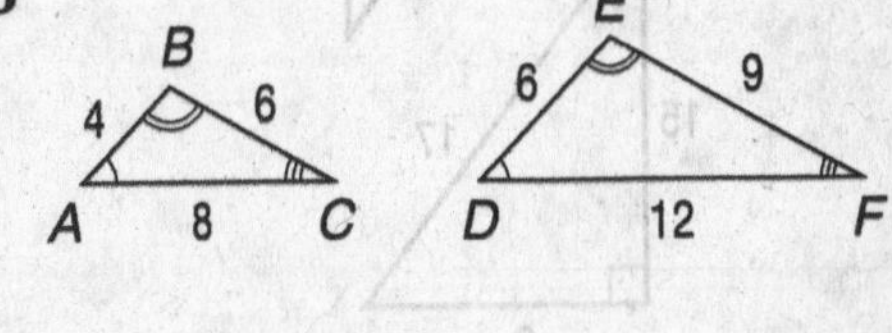

$\angle A \cong \angle D$, $\angle B \cong \angle E$, $\angle C \cong \angle F$,

$\frac{AB}{DE} = \frac{4}{6}$ or $\frac{2}{3}$, $\frac{BC}{EF} = \frac{6}{9}$ or $\frac{2}{3}$, $\frac{AC}{DF} = \frac{8}{12}$ or $\frac{2}{3}$

The corresponding angles are congruent, and the corresponding sides are proportional.

Thus, $\triangle ABC$ is similar to $\triangle DEF$.

Example 2 **Given that polygon $KLMN \sim$ polygon $PQRS$, write a proportion to find the measure of $\overline{PQ}$. Then solve.**

The ratio of corresponding sides from polygon $KLMN$ to polygon $PQRS$ is $\frac{4}{3}$. Write a proportion with this scale factor. Let x represent the measure of $\overline{PQ}$.

$\frac{KL}{PQ} = \frac{4}{3}$	$\overline{KL}$ corresponds to $\overline{PQ}$. The scale factor is $\frac{4}{3}$.
$\frac{5}{x} = \frac{4}{3}$	$KL = 5$ and $PQ = x$
$5 \cdot 3 = x \cdot 4$	Find the cross products.
$\frac{15}{4} = \frac{4x}{4}$	Multiply. Then divide each side by 4.
$3.75 = x$	Simplify.

Exercises

1. Determine whether the polygons below are similar. Explain your reasoning.

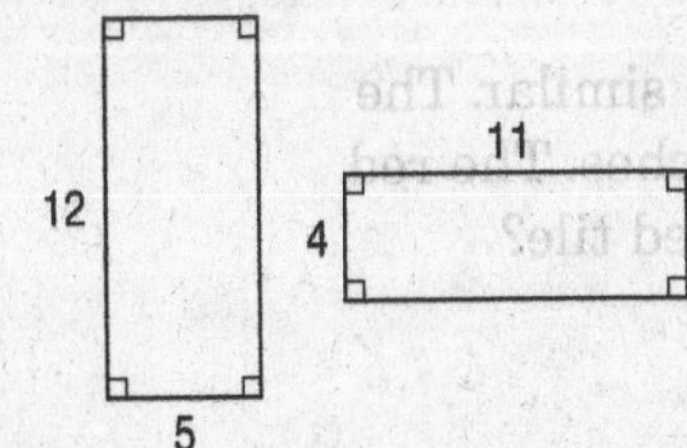

2. The triangles below are similar. Write a proportion to find each missing measure. Then solve.

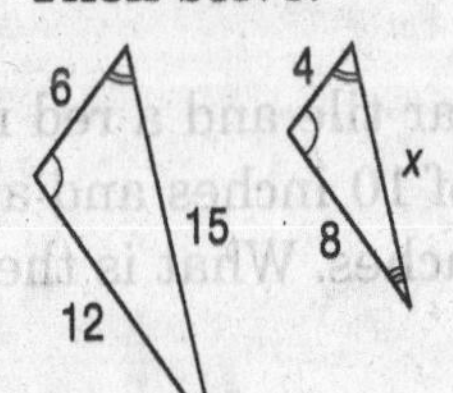

NAME ______________________ DATE __________ PERIOD _____

4-7 Practice

Similar Polygons

Determine whether each pair of polygons is similar. Explain.

1.

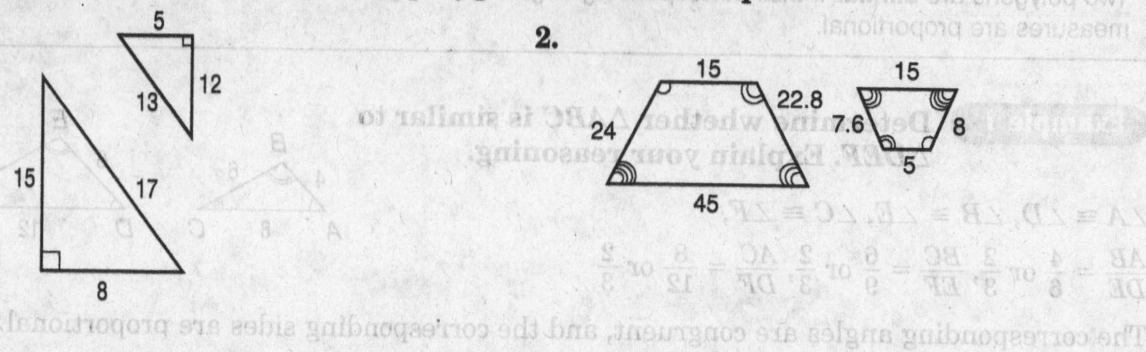

2.

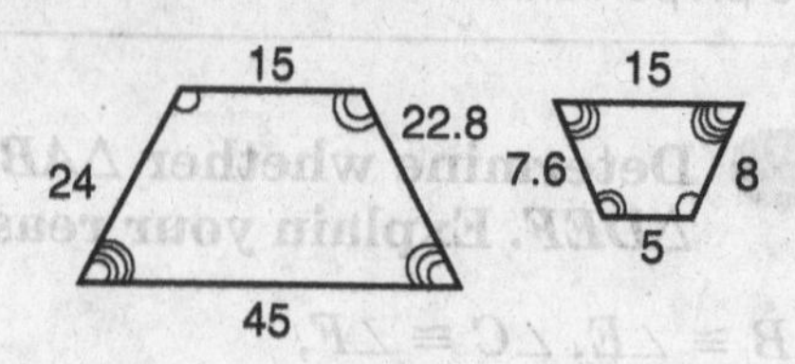

Each pair of polygons is similar. Write and solve a proportion to find each missing measure.

3.

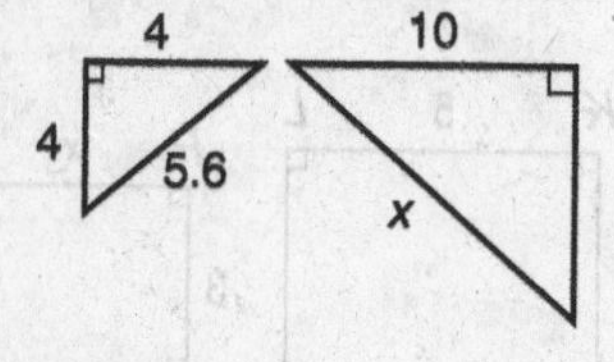

4.

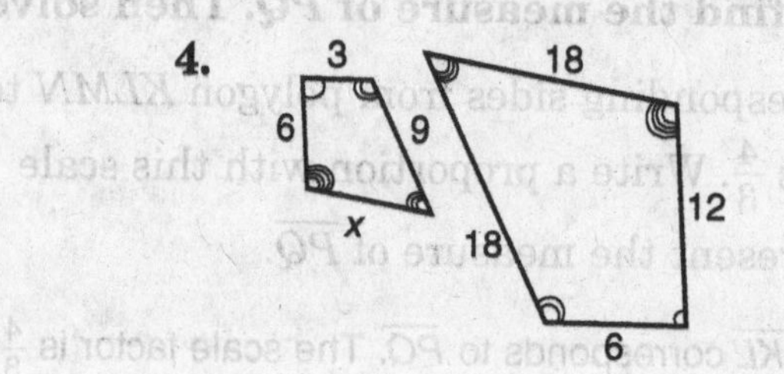

5.

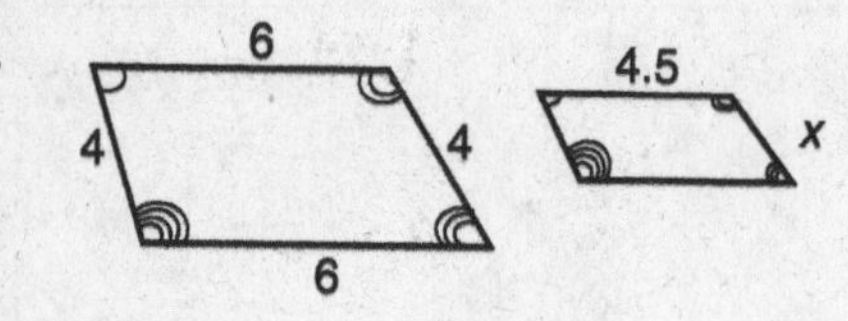

6.

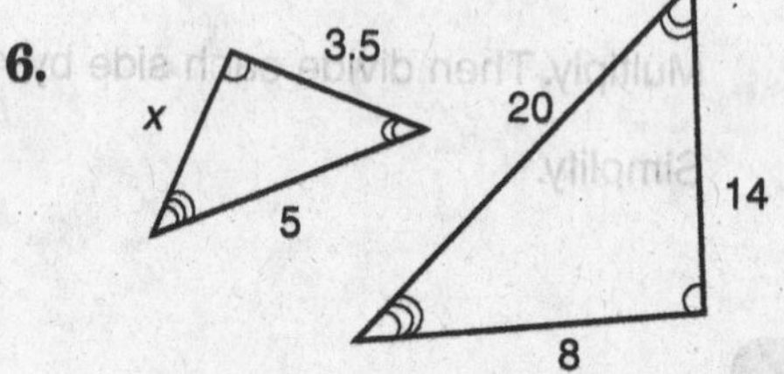

7. TILES A blue rectangular tile and a red rectangular tile are similar. The blue tile has a length of 10 inches and a perimeter of 30 inches. The red tile has a length of 6 inches. What is the perimeter of the red tile?

NAME ______________________________ DATE __________ PERIOD _____

4-8 Study Guide and Intervention

Dilations

The image produced by enlarging or reducing a figure is called a **dilation.**

Example 1 **Graph $\triangle ABC$ with vertices $A(-2, -1)$, $B(2, 3)$, and $C(2, -1)$. Then graph its image $\triangle A'B'C'$ after a dilation with a scale factor of $\frac{3}{2}$.**

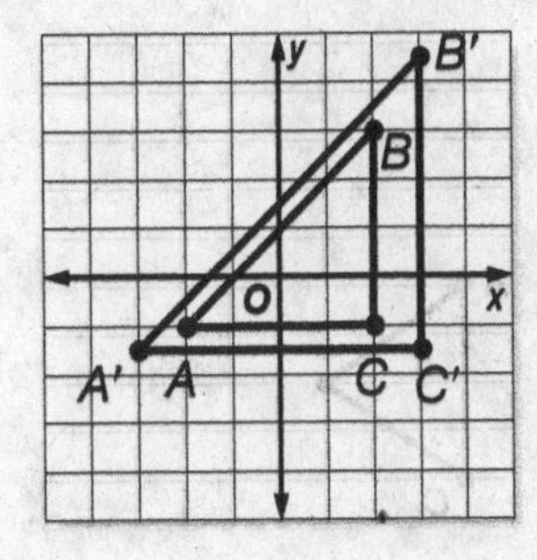

$A(-2, -1) \rightarrow \left(-2 \cdot \frac{3}{2}, -1 \cdot \frac{3}{2}\right) \rightarrow A'\left(-3, -1\frac{1}{2}\right)$

$B(2, 3) \rightarrow \left(2 \cdot \frac{3}{2}, 3 \cdot \frac{3}{2}\right) \rightarrow B'\left(3, 4\frac{1}{2}\right)$

$C(2, -1) \rightarrow \left(2 \cdot \frac{3}{2}, -1 \cdot \frac{3}{2}\right) \rightarrow C'\left(3, -1\frac{1}{2}\right)$

Example 2 **Segment $M'N'$ is a dilation of segment MN. Find the scale factor of the dilation, and classify it as an *enlargement* or a *reduction*.**

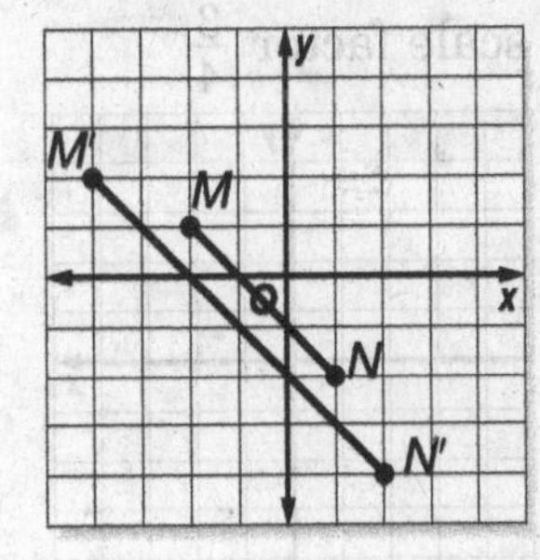

Write the ratio of the x- or y-coordinate of one vertex of the dilated figure to the x- or y-coordinate of the corresponding vertex of the original figure. Use the x-coordinates of $N(1, -2)$ and $N'(2, -4)$.

$$\frac{x\text{-coordinate of point } N'}{x\text{-coordinate of point } N} = \frac{2}{1} \text{ or } 2$$

The scale factor is 2. Since the image is larger than the original figure, the dilation is an enlargement.

Exercises

1. Polygon $ABCD$ has vertices $A(2, 4)$, $B(-1, 5)$, $C(-3, -5)$, and $D(3, -4)$. Find the coordinates of its image after a dilation with a scale factor of $\frac{1}{2}$. Then graph polygon $ABCD$ and its dilation.

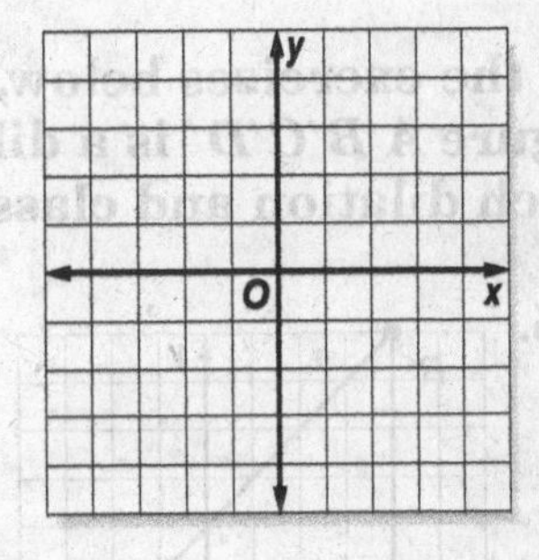

2. Segment $P'Q'$ is a dilation of segment PQ. Find the scale factor of the dilation, and classify it as an *enlargement* or a *reduction*.

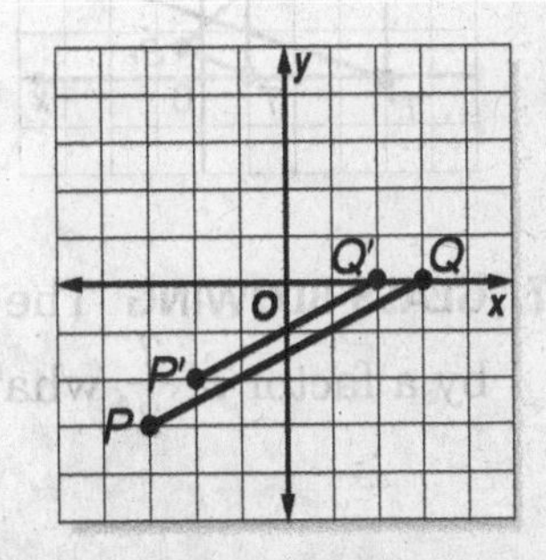

NAME ______________________ DATE __________ PERIOD ____

4-8 Practice

Dilations

Draw the image of the figure after the dilation with the given center and scale factor.

1. center: C, scale factor: 2

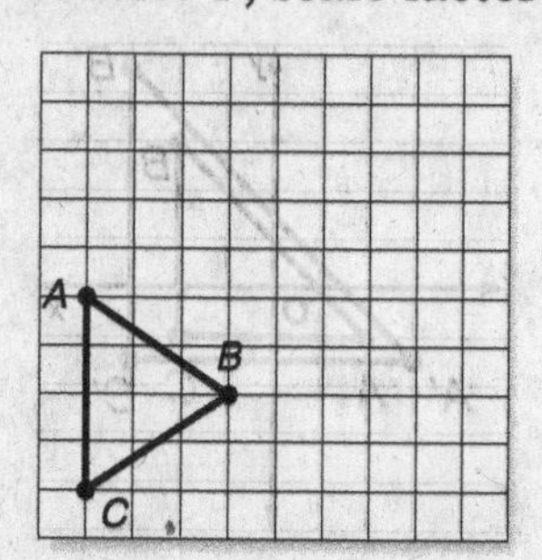

2. center: N, scale factor: $\frac{1}{2}$

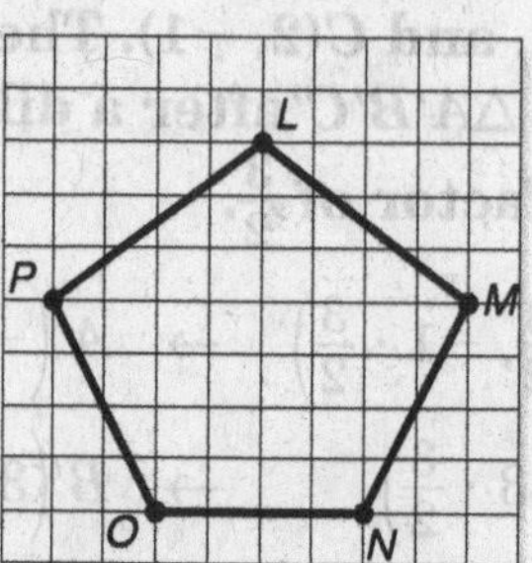

Find the coordinates of the vertices of polygon $F'G'H'J'$ after polygon $FGHJ$ is dilated using the given scale factor. Then graph polygon $FGHJ$ and polygon $F'G'H'J'$.

3. $F(-2, 2)$, $G(2, 3)$, $H(3, -2)$, $J(-1, -3)$; scale factor $\frac{3}{4}$

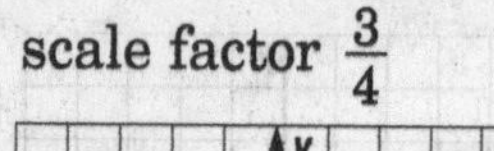

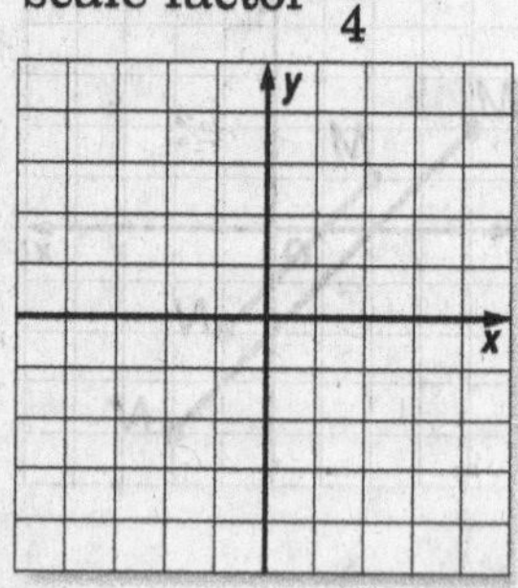

4. $F(-2, 2)$, $G(2, 4)$, $H(3, -3)$, $J(-4, -4)$; scale factor 2

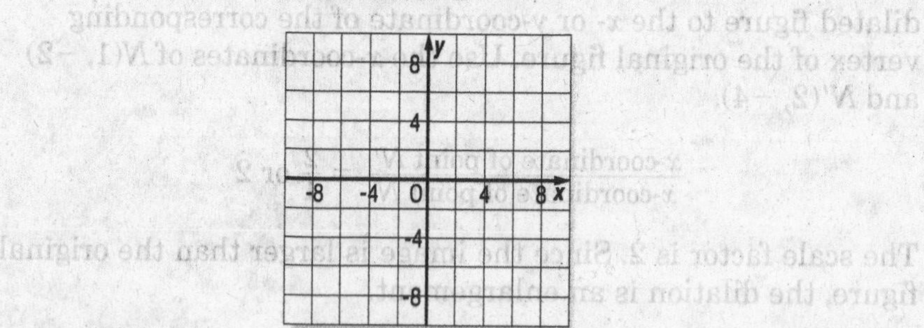

In the exercises below, figure $R'S'T'$ is a dilation of figure RST and figure $A'B'C'D'$ is a dilation of figure $ABCD$. Find the scale factor of each dilation and classify it as an *enlargement* or as a *reduction*.

5.

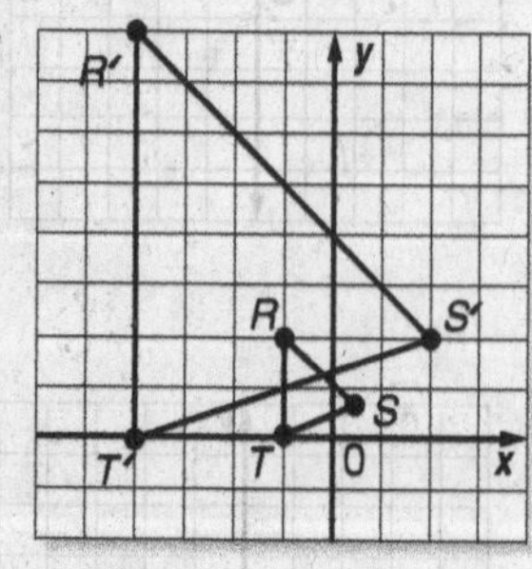

6.

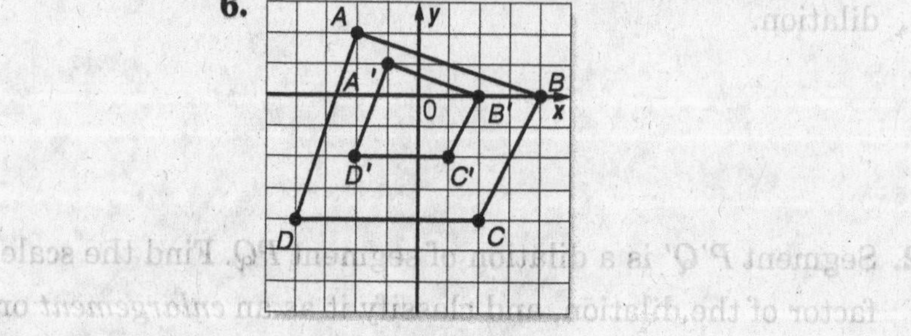

7. GLASS BLOWING The diameter of a vase is now 4 centimeters. If the diameter increases by a factor of $\frac{7}{3}$, what will be the diameter then?

4-9 Study Guide and Intervention

Indirect Measurement

Indirect measurement allows you to find distances or lengths that are difficult to measure directly using the properties of similar polygons.

Example LIGHTING **George is standing next to a lightpole in the middle of the day. George's shadow is 1.5 feet long, and the lightpole's shadow is 4.5 feet long. If George is 6 feet tall, how tall is the lightpole?**

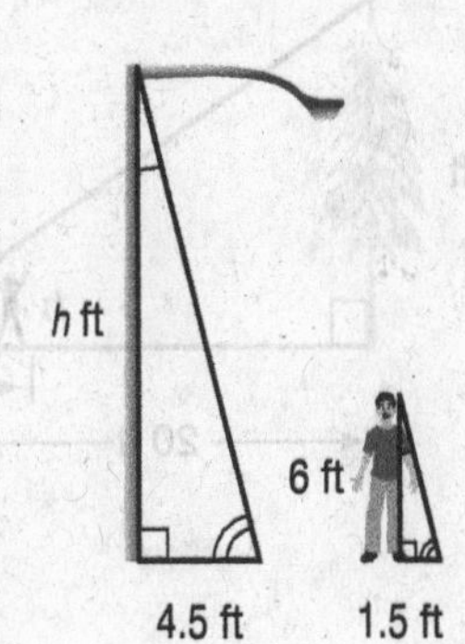

Write a proportion and solve.

George's shadow → $\frac{1.5}{4.5} = \frac{6}{h}$ ← George's height

lightpole's shadow → ← lightpole's height

$1.5 \cdot h = 4.5 \cdot 6$ **Find the cross products.**

$1.5h = 27$ **Multiply.**

$\frac{1.5h}{1.5} = \frac{27}{1.5}$ **Divide each side by 1.5.**

$h = 18$ **Simplify.**

The lightpole is 18 feet tall.

Exercises

1. **MONUMENTS** A statue casts a shadow 30 feet long. At the same time, a person who is 5 feet tall casts a shadow that is 6 feet long. How tall is the statue?

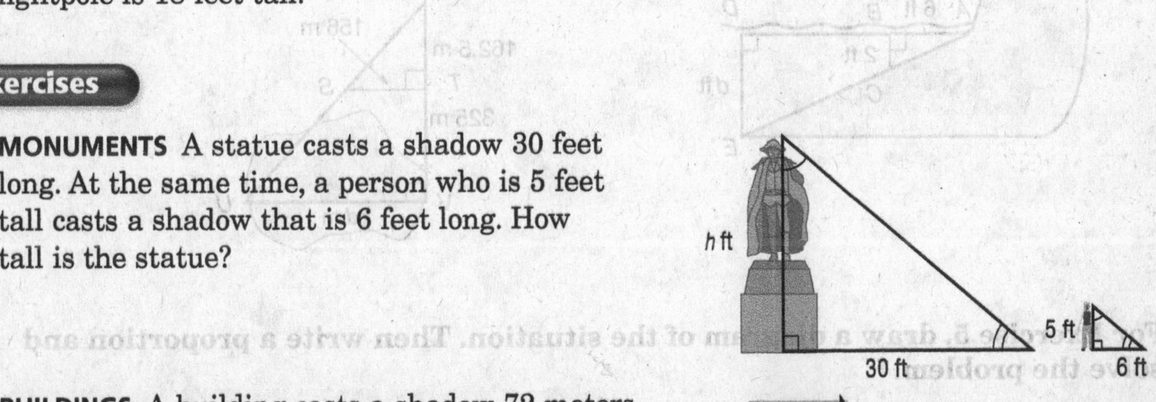

2. **BUILDINGS** A building casts a shadow 72 meters long. At the same time, a parking meter that is 1.2 meters tall casts a shadow that is 0.8 meter long. How tall is the building?

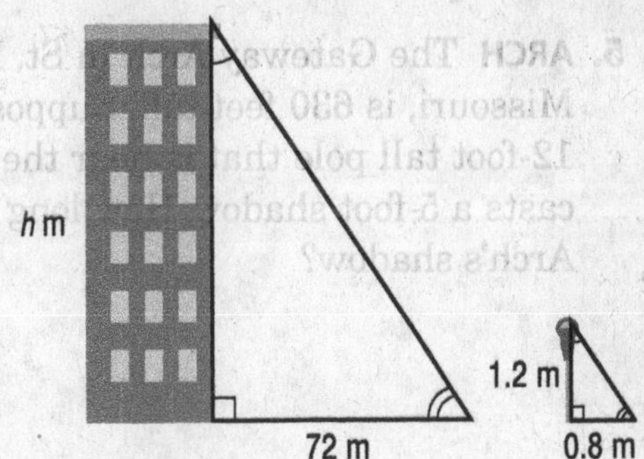

3. **SURVEYING** The two triangles shown in the figure are similar. Find the distance d across Red River.

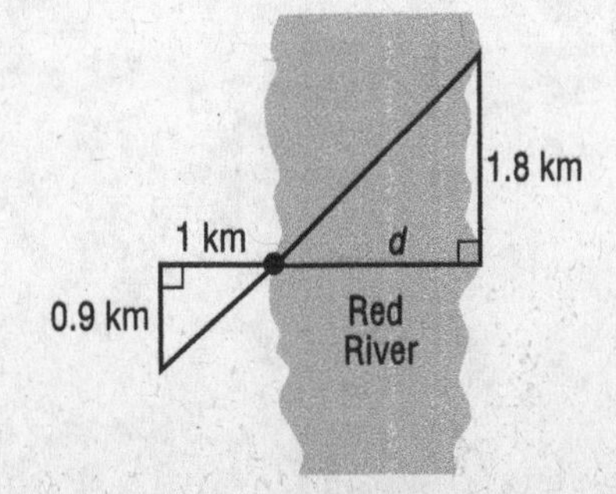

NAME ______________________ DATE __________ PERIOD _____

4-9 Practice

Indirect Measurement

In Exercises 1-4, the triangles are similar. Write a proportion and solve the problem.

1. **TREES** How tall is Yori?

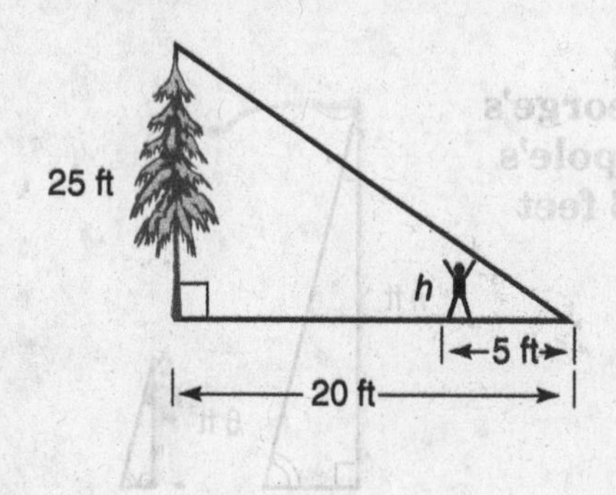

2. **TREASURE HUNT** How far is it from the hut to the gold coins?

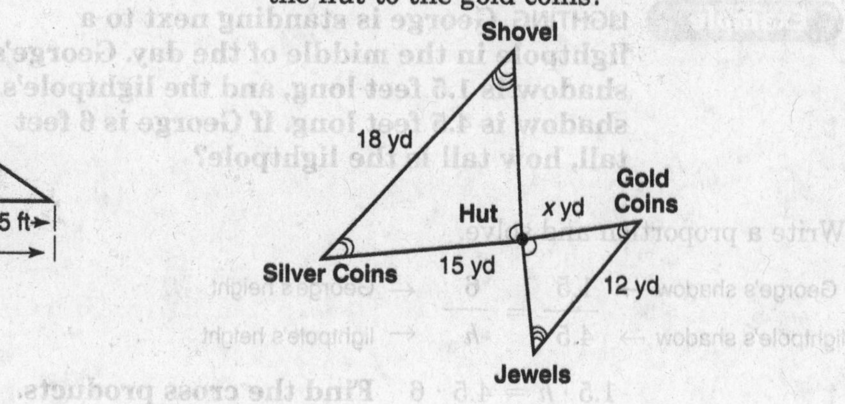

3. **LAKE** How deep is the water 31.5 feet from the shore? (Hint: $\triangle ABC \sim \triangle ADE$)

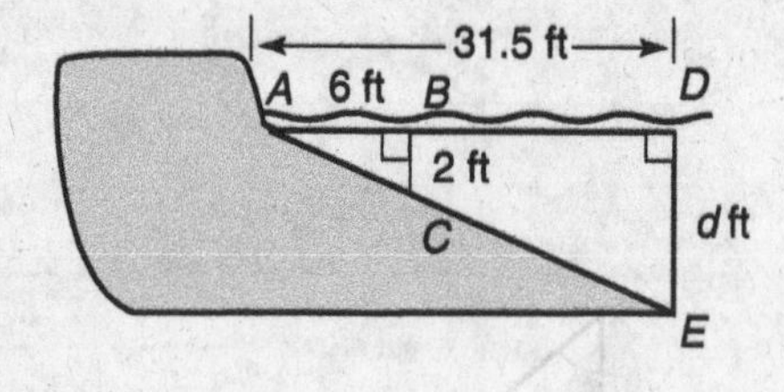

4. **SURVEYING** How far is it across the pond? (Hint: $\triangle RST \sim \triangle RUV$)

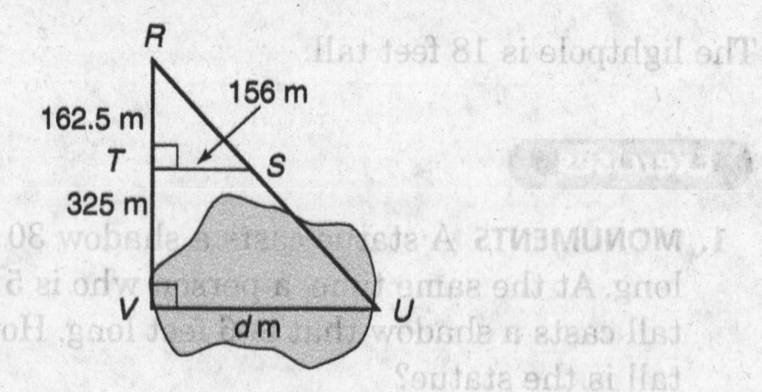

For Exercise 5, draw a diagram of the situation. Then write a proportion and solve the problem.

5. **ARCH** The Gateway Arch in St. Louis, Missouri, is 630 feet tall. Suppose a 12-foot tall pole that is near the Arch casts a 5-foot shadow. How long is the Arch's shadow?

4-10 Study Guide and Intervention

Scale Drawings and Models

Distances on a scale drawing or model are proportional to real-life distances. The **scale** is determined by the ratio of a given length on a drawing or model to its corresponding actual length.

Example 1 INTERIOR DESIGN **A designer has made a scale drawing of a living room for one of her clients. The scale of the drawing is 1 inch = $1\frac{1}{3}$ feet. On the drawing, the sofa is 6 inches long. Find the actual length of the sofa.**

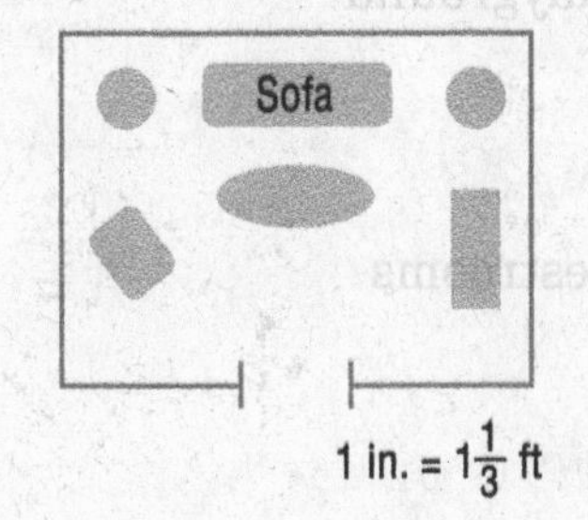

Let x represent the actual length of the sofa. Write and solve a proportion.

drawing distance → $\frac{1 \text{ in.}}{1\frac{1}{3} \text{ ft}} = \frac{6 \text{ in.}}{x \text{ ft}}$ ← drawing length
actual distance → ← actual length

$1 \cdot x = 1\frac{1}{3} \cdot 6$ Find the cross products.

$x = 8$ Simplify.

The actual length of the sofa is 8 feet.

To find the scale factor for scale drawings and models, write the ratio given by the scale in simplest form using the same units.

Example 2 **Find the scale factor for the drawing in Example 1.**

Write the ratio of 1 inch to $1\frac{1}{3}$ feet in simplest form using the same units.

$\frac{1 \text{ in.}}{1\frac{1}{3} \text{ ft}} = \frac{1 \text{ in.}}{16 \text{ in.}}$ **Convert $1\frac{1}{3}$ feet to inches.**

The scale factor is $\frac{1}{16}$ or 1:16. This means that each distance on the drawing is $\frac{1}{16}$ the actual distance.

Exercises

LANDSCAPING **Yutaka has made a scale drawing of his yard. The scale of the drawing is 1 centimeter = 0.5 meter.**

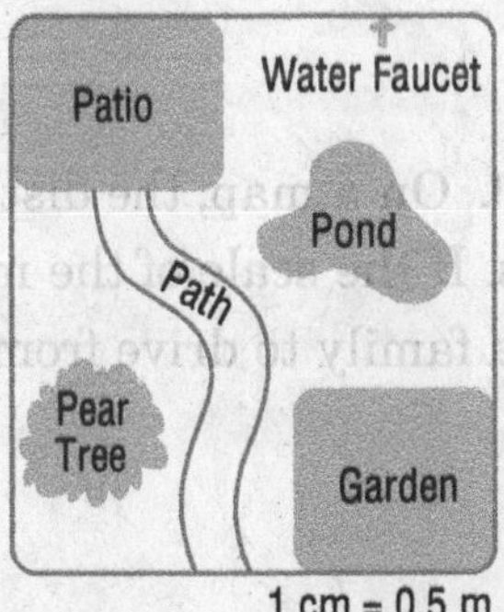

1. The length of the patio is 4.5 centimeters in the drawing. Find the actual length.

2. The actual distance between the water faucet and the pear tree is 11.2 meters. Find the corresponding distance on the drawing.

3. Find the scale factor for the drawing.

NAME ______________________ DATE ____________ PERIOD _____

4-10 Practice

Scale Drawings and Models

LANDSCAPE PLANS For Exercises 1–4, use the drawing and an inch ruler to find the actual length and width of each section of the park. Measure to the nearest eighth of an inch.

1. Playground

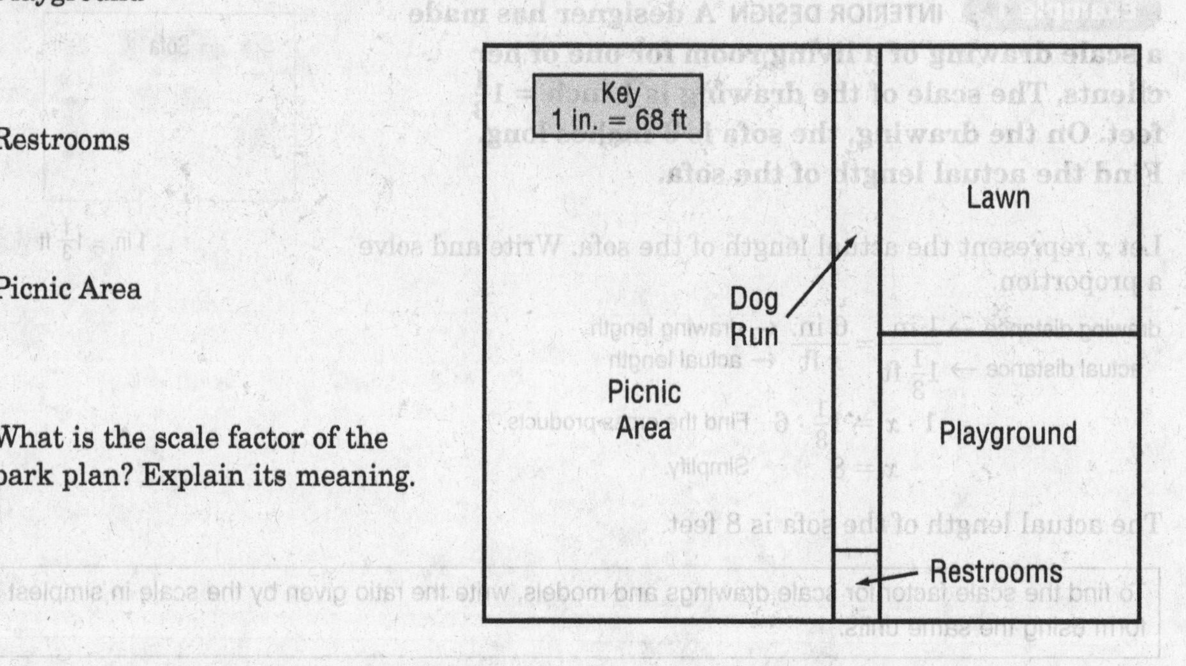

2. Restrooms

3. Picnic Area

4. What is the scale factor of the park plan? Explain its meaning.

5. **SPIDERS** The smallest spider, the *Patu marples* of Samoa, is 0.43 millimeter long. A scale model of this spider is 8 centimeters long. What is the scale of the model? What is the scale factor of the model?

6. **ANIMALS** An average adult giraffe is 18 feet tall. A newborn giraffe is about 6 feet tall. Kayla is building a model of a mother giraffe and her newborn. She wants the model to be no more than 17 inches high. Choose an appropriate scale for a model of the giraffes. Then use it to find the height of the mother and the height of the newborn giraffe.

7. **TRAVEL** On a map, the distance between Charleston and Columbia, South Carolina, is 5 inches. If the scale of the map is $\frac{7}{8}$ inch = 20 miles, about how long would it take the Garcia family to drive from Charleston to Columbia if they drove 60 miles per hour?

NAME ______________________ DATE ____________ PERIOD _____

5-1 Study Guide and Intervention

Ratios and Percents

A **percent** is a ratio that compares a number to 100. To write a fraction as a percent, find an equivalent fraction with a denominator of 100. If the denominator is a factor of 100, you can use mental math.

Examples **Write each ratio or fraction as a percent.**

1 **James made 65 out of 100 free throws.**

65 out of 100 = 65%

2 **$\frac{1}{4}$ of all high school students are not taking physics.**

$$\frac{1}{4} = \frac{25}{100} \quad (\times 25)$$

So, 1 out of 4 equals 25%.

You can express a percent as a fraction by writing it as a fraction with a denominator of 100. Then write the fraction in simplest form.

Example 3 **Write 35% as a fraction in simplest form.**

$35\% = \frac{35}{100}$ **Definition of percent.**

$= \frac{7}{20}$ **Simplify.**

So, $35\% = \frac{7}{20}$.

Exercises

Write each ratio or fraction as a percent.

1. 13 out of 100 **2.** 47 out of 100 **3.** 4:5

4. 11:20 **5.** $\frac{4}{25}$ **6.** $\frac{33}{50}$

Write each percent as a fraction in simplest form.

7. 21% **8.** 93% **9.** 10%

10. 60% **11.** 46% **12.** 88%

Lesson 5-1

NAME ______________________ DATE ____________ PERIOD _____

5-1 Practice

Ratios and Percents

Write each ratio or fraction as a percent.

1. 47 out of 100 **2.** 115 per of 100 **3.** $\frac{33}{50}$ **4.** $\frac{11}{25}$

5. 9 out of 20 **6.** 85 out of 500 **7.** 3:5 **8.** 3:10

Write each percent as a fraction in simplest form.

9. 19% **10.** 53% **11.** 30% **12.** 80%

13. 55% **14.** 48% **15.** 84% **16.** 95%

17. **GOVERNMENT** Two out of 100 U.S. Senators are from Utah. Write this ratio as a percent.

18. **ARCHITECTURE** Four out of the world's 25 tallest buildings are located in Hong Kong. Write this ratio as a percent.

19. **POPULATION** According to a recent census, the population of Montana is about 0.3% of the United States population. Write this percent as a fraction in simplest form.

20. **REASONING** Which is greatest: $\frac{3}{4}$, 19:25, or 74%? Explain your reasoning.

21. **GEOGRAPHY** Five of the 50 U.S. states border the Pacific Ocean. What percent of the U.S. states border the Pacific Ocean?

NAME ______________________ DATE __________ PERIOD _____

5-2 Study Guide and Intervention

Comparing Fractions, Decimals, and Percents

- To write a percent as a decimal, divide by 100 and remove the percent symbol.
- To write a decimal as a percent, multiply by 100 and add the percent symbol.
- To express a fraction as a percent, you can use a proportion. Alternatively, you can write the fraction as a decimal, and then express the decimal as a percent.

Example 1 **Write 56% as a decimal.**

$56\% = 56\%$ Divide by 100 and remove the percent symbol.

$= 0.56$

Example 2 **Write 0.17 as a percent.**

$0.17 = 0.17$ Multiply by 100 and add the percent symbol.

$= 17\%$

Example 3 **Write $\frac{7}{20}$ as a percent.**

Method 1 Use a proportion.

$\frac{7}{20} = \frac{x}{100}$ Write the proportion.

$7 \cdot 100 = 20 \cdot x$ Find cross products.

$700 = 20x$ Multiply.

$\frac{700}{20} = \frac{20x}{20}$ Divide each side by 20.

$35 = x$ Simplify.

So, $\frac{7}{20}$ can be written as 35%.

Method 2 Write as a decimal.

$\frac{7}{20} = 0.35$ Convert to a decimal by dividing.

$= 35\%$ Multiply by 100 and add the percent symbol.

Exercises

Write each percent as a decimal.

1. 10% **2.** 36% **3.** 82% **4.** 49.1%

Write each decimal as a percent.

5. 0.14 **6.** 0.59 **7.** 0.932 **8.** 1.07

Write each fraction as a percent.

9. $\frac{3}{4}$ **10.** $\frac{7}{10}$ **11.** $\frac{9}{16}$ **12.** $\frac{1}{40}$

Lesson 5-2

NAME ______________________________ DATE ______________ PERIOD _____

5-2 Practice

Comparing Fractions, Decimals, and Percents

Write each percent as a decimal.

1. 70% **2.** 40% **3.** 135% **4.** 369%

5. 0.9% **6.** 52.5% **7.** 8% **8.** 3%

Write each decimal as a percent.

9. 0.73 **10.** 0.84 **11.** 0.375 **12.** 0.232

13. 0.005 **14.** 0.008 **15.** 4.11 **16.** 3.52

Write each fraction as a percent.

17. $\frac{13}{25}$ **18.** $\frac{19}{20}$ **19.** $\frac{5}{4}$ **20.** $\frac{9}{5}$

21. $\frac{3}{40}$ **22.** $\frac{7}{125}$ **23.** $\frac{5}{9}$ **24.** $\frac{1}{3}$

Order each set of numbers from least to greatest.

25. $\frac{2}{5}$, 0.5, 4%, $\frac{3}{10}$ **26.** 0.6, 6%, $\frac{3}{20}$, $\frac{4}{25}$

27. 93%, 0.96, $\frac{47}{50}$, $\frac{19}{20}$ **28.** 77%, $\frac{3}{4}$, $\frac{19}{25}$, 0.73

Replace ● with <, >, or = to make a true statement.

29. $\frac{1}{200}$ ● $\frac{1}{2}$% **30.** 2.24 ● $2\frac{2}{5}$% **31.** $\frac{7}{8}$ ● $\frac{7}{8}$ %

32. TEST SCORES On a science test, Ali answered 38 of the 40 questions correctly, Jamar answered $\frac{9}{10}$ of the questions correctly, and Paco answered 92.5% of the questions correctly. Write Ali's and Jamar's scores as percents and list the students in order from the least to the highest score.

NAME ______________________ DATE ____________ PERIOD ______

5-3 Study Guide and Intervention

Algebra: The Percent Proportion

You can use a percent proportion to find a missing part, whole, or percent.

$$\frac{\text{part}}{\text{whole}} = \text{percent}$$

Example 1 **12 is what percent of 60?**

$\frac{\text{part} \rightarrow 12}{\text{whole} \rightarrow 60} = \frac{p}{100}$ } percent — Replace *a* with 12 and *b* with 60.

$12 \cdot 100 = 60 \cdot p$ — Find the cross products.

$1{,}200 = 60p$ — Multiply.

$\frac{1{,}200}{60} = \frac{60p}{60}$ — Divide each side by 60.

$20 = p$ — 12 is 20% of 60.

Example 2 **What number is 40% of 55?**

$\frac{\text{part} \rightarrow a}{\text{whole} \rightarrow 55} = \frac{40}{100}$ } percent — Replace *p* with 40 and *b* with 55.

$a \cdot 100 = 55 \cdot 40$ — Find the cross products.

$a = 22$ — Use similar steps to solve for *a*.

So, 22 is 40% of 55.

Exercises

Write a percent proportion and solve each problem. Round to the nearest tenth if necessary.

1. 3 is what percent of 10?

2. What number is 15% of 40?

3. 24 is 75% of what number?

4. 86 is what percent of 200?

5. What number is 65% of 120?

6. 52 is 13% of what number?

7. 35 is what percent of 56?

8. What number is 12.5% of 88?

9. 161 is 92% of what number?

10. 45 is what percent of 66?

11. What number is 31.5% of 200?

12. 81 is 54% of what number?

Lesson 5-3

NAME ______________________________ DATE ____________ PERIOD _____

5-3 Practice

Algebra: The Percent Proportion

Write a percent proportion and solve each problem. Round to the nearest tenth if necessary.

1. 6 is what percent of 24?

2. 125 is what percent of 375?

3. What is 20% of 80?

4. What is 14% of 440?

5. 28 is 35% of what number?

6. 63 is 63% of what number?

7. 16.24 is what percent of 14?

8. Find 350% of 49.

9. What percent of 120 is 24?

10. What percent of 84 is 6?

11. What is 7.5% of 225?

12. 9 is what percent of 660?

13. 110 is 21.1% of what number?

14. Find 6.4% of 72.

15. What percent of 160 is 1?

16. 83 is 12.5% of what number?

17. GAMES Before discarding, Carolee has 4 green cards, 3 red cards, 3 orange cards, and 1 gold card. If she discards the gold card, what percent of her remaining cards are red?

NAME ______________________ DATE __________ PERIOD ______

5-4 Study Guide and Intervention

Finding Percents Mentally

To find 1% of a number mentally, move the decimal point two places to the left. To find 10% of a number mentally, move the decimal point one place to the left.

Example 1 **Find 1% of 195.**

1% of 195 = 0.01 of 195 or 1.95

Example 2 **Find 10% of 3.9.**

10% of 3.9 = 0.1 of 3.9 or 0.39

When you compute with common percents like 50% or 25%, it may be easier to use the fraction form of the percent. It is a good idea to be familiar with the fraction form of some of the common percents.

$25\% = \frac{1}{4}$	$20\% = \frac{1}{5}$	$16\frac{2}{3}\% = \frac{1}{6}$	$12\frac{1}{2}\% = \frac{1}{8}$	$10\% = \frac{1}{10}$
$50\% = \frac{1}{2}$	$40\% = \frac{2}{5}$	$33\frac{1}{3}\% = \frac{1}{3}$	$37\frac{1}{2}\% = \frac{3}{8}$	$30\% = \frac{3}{10}$
$75\% = \frac{3}{4}$	$60\% = \frac{3}{5}$	$66\frac{2}{3}\% = \frac{2}{3}$	$62\frac{1}{2}\% = \frac{5}{8}$	$70\% = \frac{7}{10}$
	$80\% = \frac{4}{5}$	$83\frac{1}{3}\% = \frac{5}{6}$	$87\frac{1}{2}\% = \frac{7}{8}$	$90\% = \frac{9}{10}$

Example 3 **Find 25% of 68.**

25% of $68 = \frac{1}{4}$ of 68 or 17

Example 4 **Find $33\frac{1}{3}\%$ of 57.**

$33\frac{1}{3}\%$ of $57 = \frac{1}{3}$ of 57 or 19

Exercises

Compute mentally.

1. 20% of 50

2. 10% of 70

3. 50% of 34

4. 1% of 210

5. 60% of 25

6. 30% of 40

7. $66\frac{2}{3}\%$ of 33

8. $37\frac{1}{2}\%$ of 48

9. 75% of 36

10. 10% of 23

11. $83\frac{1}{3}\%$ of 24

12. 1% of 45

NAME ______________________ DATE __________ PERIOD _____

5-4 Practice

Finding Percents Mentally

Compute mentally.

1. 50% of 72 **2.** 25% of 60 **3.** $12\frac{1}{2}$% of 88 **4.** $33\frac{1}{3}$% of 84

5. 60% of 25 **6.** 20% of 150 **7.** $37\frac{1}{2}$% of 80 **8.** $83\frac{1}{3}$% of 120

9. 10% of 85 **10.** 10% of 149 **11.** 1% of 99 **12.** 1% of 78.2

13. 4% of 110 **14.** 6% of 310 **15.** 10% of 15.5 **16.** 10% of 18.4

Replace each ● with <, >, or = to make a true statement.

17. $62\frac{1}{2}$% of 40 ● 40% of 60 **18.** 10% of 85 ● 1% of 850

19. POPULATION The state of New York has a population of about 20,000,000 people. About 25% of the population of New York is under 18 years old. How many people in New York are under 18 years old?

20. LIVESTOCK In 2004, there were about 60,000,000 pigs and hogs in the United States. About 10% of the pigs and hogs were in Minnesota. How many pigs and hogs were in Minnesota in 2004?

MUSEUMS For Exercises 21–23, use the following information.

The graph shows the percents of men, women, and children visiting a modern art exhibit at a local museum. Suppose 600 people visited the exhibit.

21. How many men visited the exhibit?

22. How many women visited the exhibit?

23. How many children visited the exhibit?

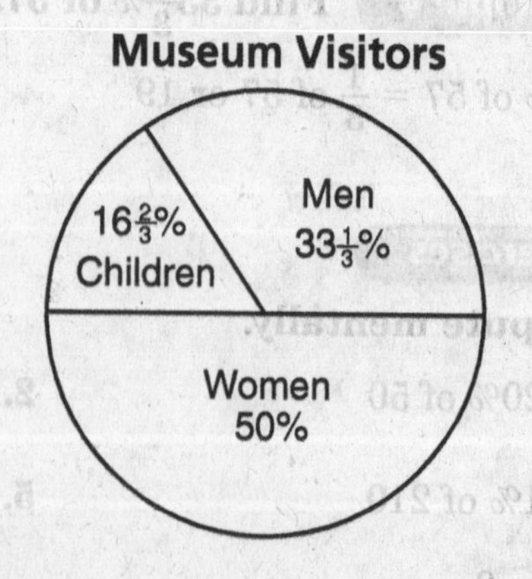

5-5 Study Guide and Intervention

Problem-Solving Investigation: Reasonable Answers

In the four-step problem-solving plan, remember that the last step is to check for reasonable answers.

Understand	• Determine what information is given in the problem and what you need to find.
Plan	• Select a strategy including a possible estimate.
Solve	• Solve the problem by carrying out your plan.
Check	• Examine your answer to see if it seems reasonable.

Example **The cost of a guitar is $300. Margaret works at the music store and can buy the guitar for 65% of the price. Will she have to pay more or less than $200?**

Understand You know the cost of the guitar. Margaret can buy the guitar for 65% of the price. You want to know if the guitar will cost more or less than $200.

Plan Find a close estimate. 65% is close to 66.66% or $\frac{2}{3}$. Multiply the cost by the estimate.

$\$300 \times \frac{2}{3} = \$200$

Solve Think. $\$300 \times \frac{2}{3} = \$200$. 65% is less than 66.66%, so she will have to pay less than $200.

Check Find 65% of $300. $\$300.00 \times .65 = \$195$.

$\$195 < \$200.00$ The answer is reasonable.

Exercises

For Exercises 1–5, determine a reasonable answer.

1. **JOBS** Maxine is paid $9.25 an hour to work at the bookstore. If she is saving to buy a new video game system that costs $360, will she have to work 30, 40, or 50 hours?

2. **MONEY** Jeff brings $120 to purchase winter clothes. He buys a coat for $57.36. He wants to purchase a pair of jeans for $28.95 and a pair of boots for $54.98. Does he have enough money with him to make these two purchases?

3. **SURVEY** In a recent survey, 56% of students at Trenton Middle School work at part-time jobs during the school year. If there are 1,378 students in the school, is 550, 650, or 750 a reasonable estimate for the number of students who work part time during the school year?

4. **SHOPPING** Byron took $80 to the mall to buy gifts. He spent $28.73 on a video game. He wants to purchase a book for $13.89 and a laptop bag for $39.99. Does he have enough money with him to make these two purchases?

5. **ATTENDANCE** There are 1,200 students at Hillsboro Middle School. If 43% of the students attend an exhibit given by the art department, would the number of students who attended be 924, 516, or 430?

NAME ______________________ DATE __________ PERIOD _____

5-5 Practice

Problem-Solving Investigation: Reasonable Answers

Mixed Problem Solving

Use the reasonable answer strategy to solve Exercises 1 and 2.

1. **POPULATION** About 9.5% of the population of New Mexico is Native American. If the population of New Mexico is 1,874,614, would the number of Native Americans living in New Mexico be about 180,000, 360,000, or 900,000?

2. **HOMES** Mr. and Mrs. Whatley want to buy a new home for $245,000. The bank requires 20% of the price of the home as a down payment for the loan. Should the Whatleys plan to pay $5,000, $25,000, or $50,000 as the down payment?

Use any strategy to solve Exercises 3–6. Some strategies are shown below.

PROBLEM-SOLVING STRATEGIES
• Work backward.   • Look for a pattern.   • Draw a diagram.

3. **SPORTS** Three teams participating in a track meet have 25 members, 29 members, and 33 members. The coach of the hosting team wants to have three bottles of water for each athlete. If each case of water contains 24 bottles, should the coach buy 4, 12, or 20 cases of water?

4. **MONEY** After Latoya gave 35% of her allowance to her brother and 25% of her allowance to her sister, she had $12 left. How much was Latoya's allowance?

5. **ELECTIONS** A county with 31,500 registered voters is buying new voting machines. State law requires that the county have one polling place for every 750 registered voters and 4 voting machines per polling place. How many new voting machines should the county order?

6. **GEOMETRY** Brandon is drawing a rectangle similar to the one below except that each side of his rectangle is $2\frac{1}{2}$ times longer. Find the area of Brandon's rectangle.

8 cm

2.4 cm

5-6 Study Guide and Intervention

Percent and Estimation

You can use compatible numbers to estimate a percent of a number. **Compatible numbers** are two numbers that are easy to divide mentally.

Example 1 **Estimate 35% of 60.**

35% is about $33\frac{1}{3}\%$ or $\frac{1}{3}$. $\frac{1}{3}$ and 60 are compatible numbers.

$\frac{1}{3}$ of 60 is 20.

So, 35% of 60 is about 20.

Example 2 **Estimate what percent corresponds to 23 out of 59.**

$\frac{23}{59} \approx \frac{24}{60}$ or $\frac{2}{5}$ 23 is about 24, and 59 is about 60.

$\frac{2}{5} = 40\%$

So, 23 out of 59 is about 40%.

Exercises

Estimate.

1. 11% of 60

2. 24% of 36

3. 81% of 25

4. 19% of 41

5. 32% of 66

6. 67% of 44

Estimate each percent.

7. 7 out of 15

8. 6 out of 23

9. 5 out of 51

10. 8 out of 35

11. 13 out of 17

12. 17 out of 26

NAME ______________________ DATE ____________ PERIOD _____

5-6 Practice

Percent and Estimation

Estimate.

1. 39% of 80 **2.** 66% of 72 **3.** 40% of 89 **4.** 75% of 35

5. 19% of 79 **6.** 72% of 51 **7.** 53% of 199 **8.** 23% of 162

9. 48.5% of 151 **10.** 76.5% of 303 **11.** 148% of 69 **12.** 226% of 81

Estimate each percent.

13. 8 out of 37 **14.** 4 out of 19 **15.** 10 out of 21 **16.** 29 out of 90

17. 7 out of 9 **18.** 29 out of 40 **19.** 9 out of 31 **20.** 11 out of 59

21. ANALYZE TABLES The table gives the land area of one county in each state and the land area of the entire state. Estimate the percent of the land area of each state that is in the county. Then determine which county has the greatest percent of its state's land area. Round to the nearest tenth if necessary.

County	Land Area of County (square miles)	Land Area of Entire State (square miles)
Kent County, MD	279	9,774
Marion County, SC	489	30,109
Newport County, RI	104	1,045

Source: U.S. Census Bureau

5-7 Study Guide and Intervention

Algebra: The Percent Equation

A **percent equation** is an equivalent form of a percent proportion in which the percent is written as a decimal.

part = percent · whole

Example 1 **Find 22% of 245.**

The percent is 22%, and the whole is 245. Let n represent the part.

$n = 0.22(245)$ Write 22% as the decimal 0.22.
$n = 53.9$ Simplify.

So, 22% of 245 is 53.9.

Example 2 **600 is what percent of 750?**

The part is 600, and the whole is 750. Let n represent the percent.

$600 = n(750)$ Write the equation.

$\frac{600}{750} = \frac{750n}{750}$ Divide each side by 750.

$0.8 = n$ Simplify.

Since 0.8 = 80%, 600 is 80% of 750.

Example 3 **45 is 90% of what number?**

The part is 45, and the percent is 90%. Let n represent the whole.

$45 = 0.90 \cdot n$ Write 90% as the decimal 0.90.

$\frac{45}{0.90} = \frac{0.90n}{0.90}$ Divide each side by 0.90.

$50 = n$ Simplify.

So, 45 is 90% of 50.

Exercises

Solve each problem using the percent equation.

1. Find 30% of 70.
2. What is 80% of 65?
3. What percent of 56 is 14?
4. 36 is what percent of 40?
5. 80 is 40% of what number?
6. 65% of what number is 78?
7. What percent of 2,000 is 8?
8. 12 is what percent of 4,000?
9. What percent of 3,000 is 18?
10. What is 110% of 80?
11. Find 180% of 160.
12. 4% of what number is 11?

NAME ______________________ DATE __________ PERIOD _____

5-7 Practice

Algebra: The Percent Equation

Solve each problem using a percent equation.

1. Find 80% of 40.

2. What is 30% of 70?

3. What percent of 80 is 32?

4. 36 is what percent of 120?

5. 35% of what number is 84?

6. 95 is 50% of what number?

7. What number is 18% of 72?

8. Find 32% of 96.

9. 8 is what percent of 4,000?

10. What percent of 6,000 is 15?

11. 4% of what number is 7?

12. 85 is 10% of what number?

13. Find $3\frac{1}{2}\%$ of 250.

14. What is $7\frac{1}{4}\%$ of 56?

15. 560 is what percent of 420?

16. $2\frac{1}{5}\%$ of what number is 44?

17. **VIDEO GAMES** A video game costs $55. If 7.5% sales tax is added, what is the total cost of the video game?

18. **FOOTBALL** In the 2006 Super Bowl, Pittsburgh and Seattle each scored 7 points in the 4th quarter. Which team scored the higher percentage of their final score in the 4th quarter?

2006 Super Bowl

Team	Final Score
Pittsburgh	21
Seattle	10

NAME ______________________________ DATE ____________ PERIOD _____

5-8 Study Guide and Intervention

Percent of Change

To find the percent of change, first find the amount of change. Then find the ratio of that amount to the original amount, and write the ratio as a percent.

Example **Two months ago, the bicycle shop sold 50 bicycles. Last month, 55 bicycles were sold. Find the percent of change. State whether the percent of change is an *increase* or a *decrease*.**

Step 1 Subtract to find the amount of change.

$55 - 50 = 5$

Step 2 Write a ratio that compares the amount of change to the original number of bicycles.

Step 3 Write the ratio as a percent.

$\text{percent of change} = \dfrac{\text{amount of change}}{\text{original amount}}$ Definition of percent of change

$= \dfrac{5}{50}$ The amount of change is 5. The original amount is 50.

$= 0.1 \text{ or } 10\%$ Divide. Write as a percent.

The percent of change is 10%. Since the new amount is greater than the original, it is a percent of increase.

Exercises

Find each percent of change. Round to the nearest tenth of a percent if necessary. State whether the percent of change is an *increase* or a *decrease*.

1. original: 4
new: 5

2. original: 10
new: 13

3. original: 15
new: 12

4. original: 30
new: 18

5. original: 60
new: 63

6. original: 160
new: 136

7. original: 77
new: 105

8. original: 96
new: 59

Lesson 5-8

NAME ______________________________ DATE ____________ PERIOD _____

5-8 Practice

Percent of Change

Find each percent of change. Round to the nearest tenth if necessary. State whether the percent of change is an *increase* or *a decrease*.

1. original: 8 points
new: 10 points

2. original: 45 inches
new: 48 inches

3. original: $60
new: $48

4. original: $750
new: $690

5. original: 25 miles
new: 36 miles

6. original: 12 fouls
new: 8 fouls

Find the selling price for each item given the cost to the store and the percent of markup.

7. backpack: $14, 40% markup

8. soccer ball: $22, 35% markup

9. music CD: $9, 45% markup

10. sweatshirt: $27, 20% markup

Find the sale price of each item to the nearest cent.

11. book: $29, 25% off

12. sofa: $975, 30% off

13. jeans: $34.95, 40% off

14. stereo: $459.99, 15% off

Find each percent of change. Round to the nearest tenth if necessary.

15. What is the percent of markup on a $120 cell phone that sells for $149?

16. Find the percent of markup on a $50 pair of shoes that sells for $60.

17. Find the percent of discount on a $45 jacket that regularly sells for $75.

18. What is the percent of discount on a $290 television that regularly sells for $349?

NAME ______________________ DATE ____________ PERIOD _____

5-9 Study Guide and Intervention

Simple Interest

To find simple interest, use the formula $I = prt$. Interest I is the amount of money paid or earned. Principal p is the amount of money invested or borrowed. Rate r is the annual interest rate. Time t is the time in years.

Example 1 **Find the simple interest for $600 invested at 8.5% for 6 months.**

Notice the time is given in months. Six months is $\frac{6}{12}$ or $\frac{1}{2}$ year.

$I = prt$ — Write the simple interest formula.

$I = 600 \cdot 0.085 \cdot \frac{1}{2}$ — Replace p with 600, r with 0.085, and t with $\frac{1}{2}$.

$I = 25.50$ — Simplify.

The simple interest is $25.50.

Example 2 **Find the total amount in an account where $136 is invested at 7.5% for 2 years.**

$I = prt$ — Write the simple interest formula.

$I = 136 \cdot 0.075 \cdot 2$ — Replace p with 136, r with 0.075, and t with 2.

$I = 20.40$ — Simplify.

The simple interest is $20.40. The amount in the account is $136 + $20.40 = $156.40.

Exercises

Find the simple interest to the nearest cent.

1. $300 at 5% for 2 years
2. $650 at 8% for 3 years
3. $575 at 4.5% for 4 years
4. $735 at 7% for $2\frac{1}{2}$ years
5. $1,665 at 6.75% for 3 years
6. $2,105 at 11% for $1\frac{3}{4}$ years

Find the total amount in each account to the nearest cent.

7. $250 at 5% for 3 years
8. $425 at 6% for 2 years
9. $945 at 7.25% for 4 years
10. $1,250 at 7.4% for $2\frac{1}{4}$ years
11. $2,680 at 9.1% for $1\frac{3}{4}$ years
12. $4,205 at 4.5% for $3\frac{1}{2}$ years

Lesson 5-9

NAME ______________________ DATE ____________ PERIOD _____

5-9 Practice

Simple Interest

Find the simple interest to the nearest cent.

1. \$350 at 5% for 4 years

2. \$750 at 6.5% for 3 years

3. \$925 at 4.75% for 3 months

4. \$2,050 at 7.65% for 36 months

5. \$1,000 at $5\frac{1}{2}$% for 18 months

6. \$8,500 at $10\frac{3}{8}$% for $2\frac{1}{2}$ years

Find the total amount in each account to the nearest cent, assuming simple interest.

7. \$1,500 at 6% for 5 years

8. \$4,010 at 5.2% for 4 years

9. \$925 at 7.25% for 6 months

10. \$850 at 8.38% for 8 months

11. \$16,000 at $3\frac{1}{4}$% for 42 months

12. \$3,200 at $6\frac{2}{3}$% for $5\frac{1}{2}$ years

13. COINS The value of a rare coin increased in value from \$350 to \$420 in only 6 months. Find the simple interest rate for the value of the coin.

14. HOUSING Mrs. Landry bought a house for \$35,000 in 1975. She sold the house for \$161,000 in 2005. Find the simple interest rate for the value of the house.

15. CARS Brent's older brother took out a 4-year loan for \$16,000 to buy a car. If the simple interest rate was 8%, how much total will he pay for the car including interest?

16. MORTGAGE The Fongs need to borrow \$200,000 to purchase a home. A local bank is offering a 20-year mortgage at 6.75% interest, while an online lender is offering a 30-year mortgage at 5.25% interest. Assuming simple interest, which loan will result in the lower interest?

NAME ______________________________ DATE ____________ PERIOD ______

6-1 Study Guide and Intervention

Line and Angle Relationships

Angle Relationships		
Vertical Angles	**Complementary Angles**	**Supplementary Angles**
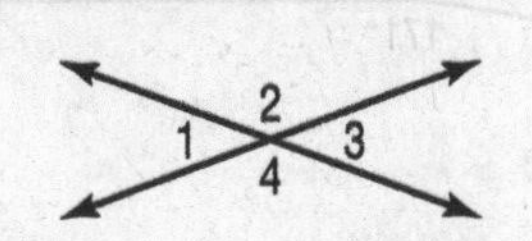	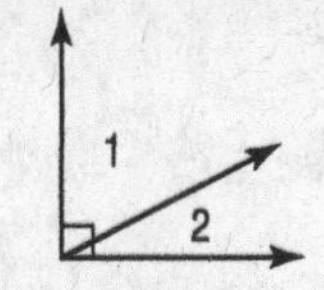	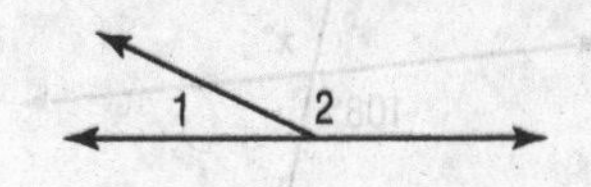
$m\angle 1 = m\angle 3$ $m\angle 2 = m\angle 4$	$m\angle 1 + m\angle 2 = 90°$	$m\angle 1 + m\angle 2 = 180°$

Points, Lines, and Planes		
Parallel Lines	**Perpendicular Lines**	**Transversal**
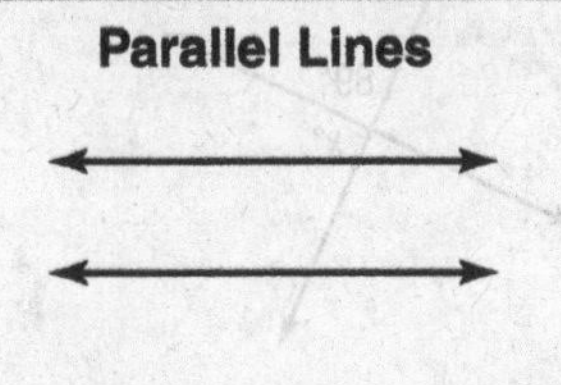	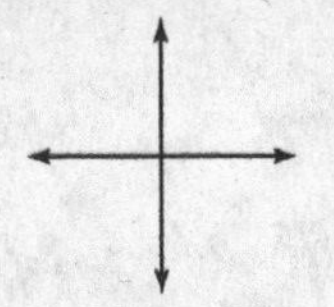	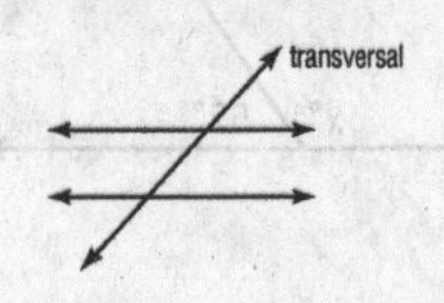

Example **Find the value of x.**

The angles are opposite each other and formed by intersecting lines, so they are vertical angles. Vertical angles are congruent.

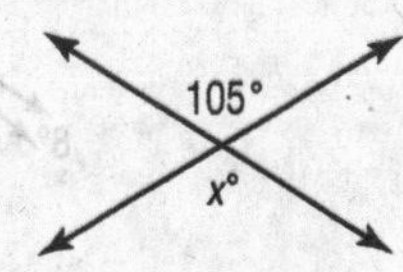

Exercises

Find the value of x in each figure.

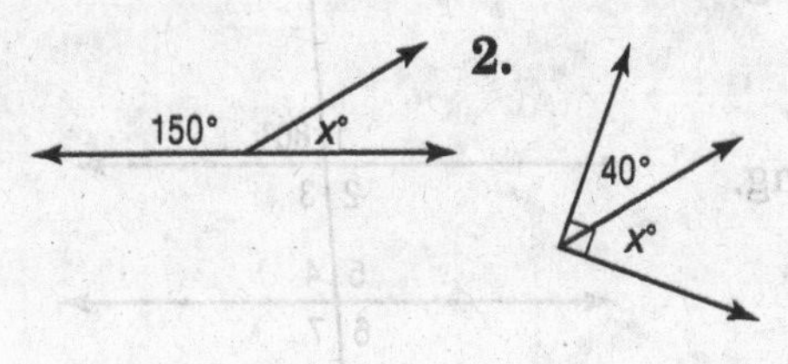

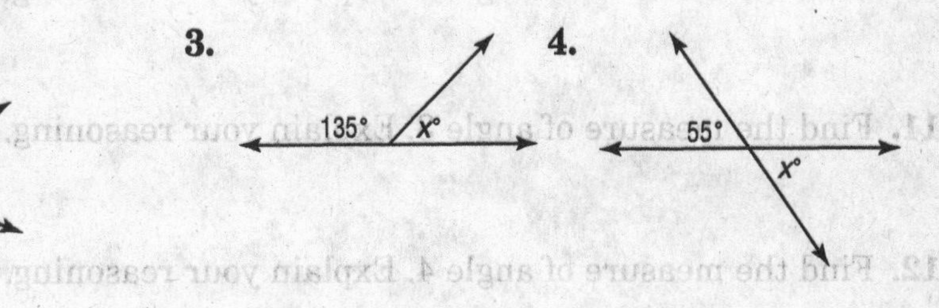

For Questions 5 and 6, use the figure at the right.

5. Find the measure of angle 2. Explain your reasoning.

6. Find the measure of angle 4. Explain your reasoning.

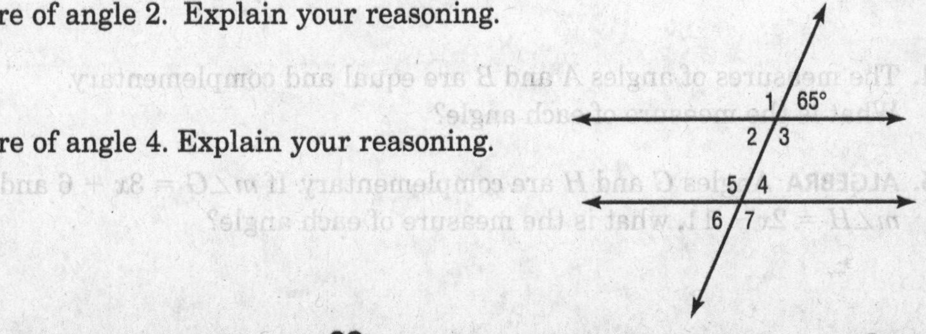

NAME ______________________ DATE __________ PERIOD _____

6-1 Practice

Line and Angle Relationships

Find the value of *x* in each figure.

1.

2.

3.

4.

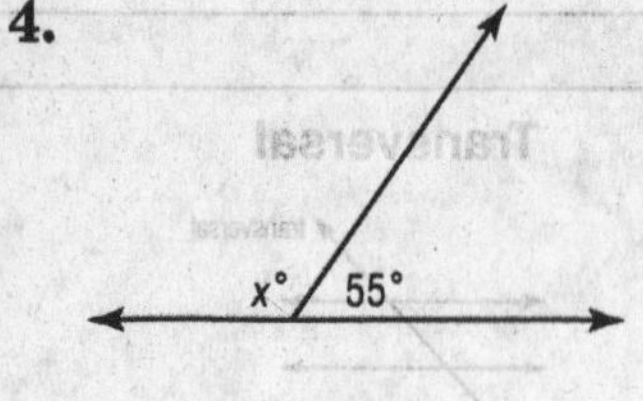

5.

6.

7.

8.

9.

Use the figure at the right to answer 10–13.

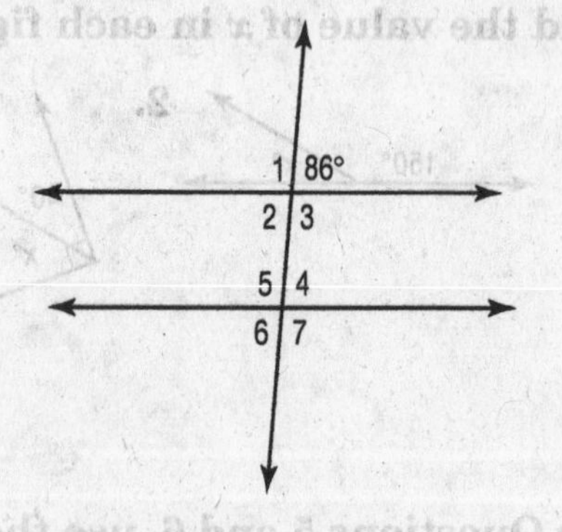

10. Find the measure of angle 2. Explain your reasoning.

11. Find the measure of angle 3. Explain your reasoning.

12. Find the measure of angle 4. Explain your reasoning.

13. Find the measure of angle 6. Explain your reasoning.

14. The measures of angles A and B are equal and complementary. What is the measure of each angle?

15. **ALGEBRA** Angles G and H are complementary. If $m\angle G = 3x + 6$ and $m\angle H = 2x - 11$, what is the measure of each angle?

6-2 Study Guide and Intervention

Problem-Solving Investigation: Use Logical Reasoning

You may need to use logical reasoning to solve some problems.

Understand	• Determine what information is given in the problem and what you need to find.
Plan	• Select a strategy including a possible estimate.
Solve	• Solve the problem by carrying out your plan.
Check	• Examine your answer to see if it seems reasonable.

Example **A plane figure has four sides. The figure has only two congruent sides and two pairs of congruent angles. Is the figure a square, rectangle, parallelogram, rhombus, or trapezoid? Did you use *deductive* or *inductive* reasoning?**

Understand We know that a plane figure has four sides and the figure has only two congruent sides and two pairs of congruent angles. We need to see if the figure is a square, rectangle, parallelogram, rhombus, or trapezoid.

Plan Let's look at the characteristics of these different figures.
A square or rhombus has *four* congruent sides.
The figure is not a square or a rhombus.
A rectangle or parallelogram has *two* pairs of congruent sides. The figure is not a rectangle or a parallelogram.

Solve An isosceles trapezoid can have two congruent sides and two pairs of congruent angles. The figure could be a trapezoid.

Check Since all choices but the trapezoid were eliminated, the figure is a trapezoid. Because you used existing rules about four-sided figures to make a decision, you used deductive reasoning.

Exercises

For Exercises 1–3, solve each problem using logical reasoning.

1. **GEOMETRY** Jennifer draws a square on a piece of paper and uses a ruler to draw one line through the square to create two shapes. What is the maximum number of sides that either of these shapes can have, and how would the line have to be drawn to create it?

2. **MODELS** You have 30 toothpicks. You can create two adjacent squares using 7 toothpicks if the adjacent square shares a toothpick for the side between them. How many total squares could be created this way with 30 toothpicks, if the squares are formed in a row?

3. **AGES** You and your grandfather have a combined age of 84 years. If your grandfather is 6 times as old as you are, how old are you? Explain.

6-2 Practice

Problem-Solving Investigation: Use Logical Reasoning

Mixed Problem Solving

For Exercises 1 and 2, solve each problem using logical reasoning.

1. **NUMBER SENSE** Simplify each product of powers. Then use logical reasoning to simplify $10^4 \times 0.1^4$, $10^5 \times 0.1^5$, and $10^{12} \times 0.1^{12}$.

Product of Powers	Simplified Form
$10^2 \times 0.1^2$	
$10^3 \times 0.1^3$	
$10^7 \times 0.1^7$	

2. **MEASUREMENT** You have a pen that is 6 inches long and a pencil that is 7 inches long. Explain how you can use the pen and pencil to draw a line segment that is 3 inches long.

Use any strategy to solve Exercises 3–6. Some strategies are shown below.

PROBLEM-SOLVING STRATEGIES
• Look for a pattern.
• Draw a diagram.
• Use logical reasoning.

3. **SPORTS** At the end of a baseball game, the winning team had three more runs than their opponents. If they had scored 1 more run, they would have had twice as many as their opponents. How many runs did each team have?

4. **SHOPPING** Brittany bought five items at the grocery store for her mother. From the given clues, list the items from least expensive to most expensive.
   - The peanut butter cost less than the sliced turkey.
   - The sliced turkey cost half as much as the birthday cake.
   - The peanut butter cost $0.20 more than the milk.
   - The price of the lettuce was 40% of the price of the milk.

5. **SOLAR SYSTEM** Jupiter is the largest planet in the solar system with a diameter of 88,736 miles. Saturn is the second largest planet with a diameter of 74,978 miles. How much greater is the diameter of Jupiter than the diameter of Saturn?

6. **TRAVEL** Mr. Bradley often flies from Chicago to San Francisco and back again, a total distance of 3,716 miles. If he made this trip 25 times last year, find the total distance Mr. Bradley traveled on these trips.

6-3 Study Guide and Intervention

Polygons and Angles

An **interior angle** is an angle with sides that are adjacent sides of the polygon. A **regular polygon** is a polygon whose sides and angles are congruent.

Example 1 **Find the sum of the measures of the interior angles of a tricontagon, which is a 30-sided polygon.**

$S = (n - 2)180°$	Write an equation.
$S = (30 - 2)180°$	Replace n with 30. Subtract.
$S = (28)180°$	Multiply.
$S = 5{,}040°$	

The sum of the measures of the interior angles of a tricontagon is 5,040°.

Example 2 **The defense department of the United States has its headquarters in a building called the Pentagon because it is shaped like a regular pentagon. What is the measure of an interior angle of a regular pentagon?**

$S = (n - 2)180°$	Write an equation.
$S = (5 - 2)180°$	Replace n with 5. Subtract.
$S = (3)180°$	Multiply.
$S = 540°$	
$540° \div 5 = 108°$	Divide by the number of interior angles to find the measure of one angle.

The measure of one interior angle of a regular pentagon is 108°.

Exercises

For Exercises 1–6, find the sum of the measures of the interior angles of the given polygon.

1. nonagon (9-sided)

2. 14-gon

3. 16-gon

4. hendecagon (11-sided)

5. 25-gon

6. 42-gon

For Exercises 7–12, find the measure of one interior angle of the given regular polygon. Round to the nearest hundredth if necessary.

7. hexagon

8. 15-gon

9. 22-gon

10. icosagon (20-sided)

11. 38-gon

12. pentacontagon (50-sided)

Lesson 6-3

NAME ______________________ DATE ____________ PERIOD _____

6-3 Practice

Polygons and Angles

Find the sum of the measures of the interior angles of each polygon.

1. 13-gon **2.** 16-gon **3.** 17-gon

4. 18-gon **5.** 20-gon **6.** 25-gon

Find the measure of one interior angle in each regular polygon. Round to the nearest tenth if necessary.

7. pentagon **8.** hexagon **9.** 24-gon

ALGEBRA For Exercises 10 and 11, determine the angle measures in each polygon.

10.

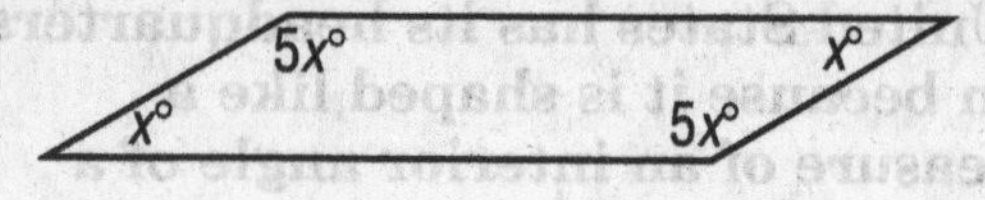

11.

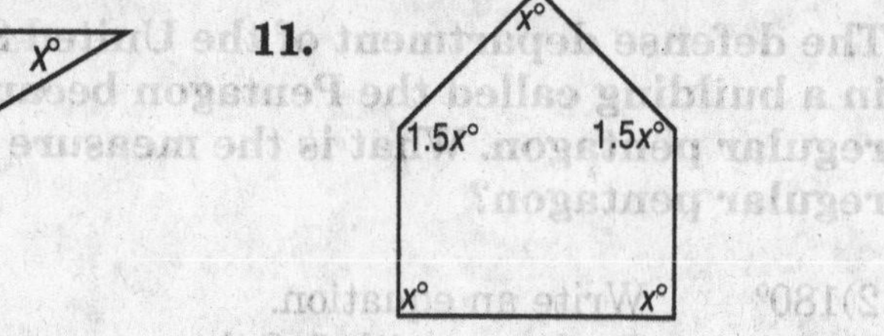

12. FLOORING A floor is tiled with a pattern consisting of regular octagons and squares as shown. Find the measure of each angle at the circled vertex. Then find the sum of the angles.

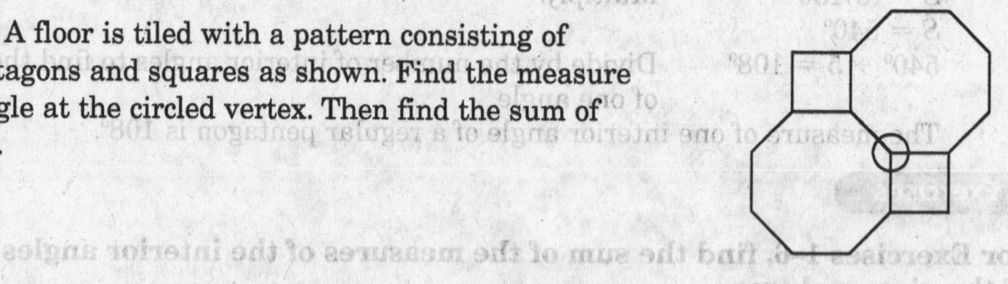

13. ART Jose is laying out a pattern for a stained glass window. So far he has placed the 13 regular polygons shown. Find the measure of each angle at the circled vertex. Then find the sum of the angles.

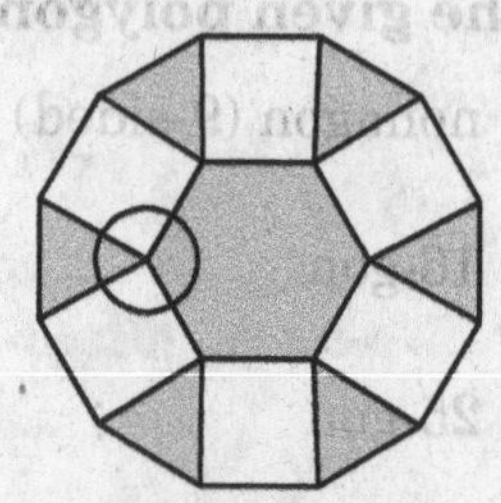

14. REASONING Vanessa's mother made a quilt using a pattern of repeating regular hexagons as shown. Will Vanessa be able to make a similar quilt with a pattern of repeating regular pentagons? Explain your reasoning.

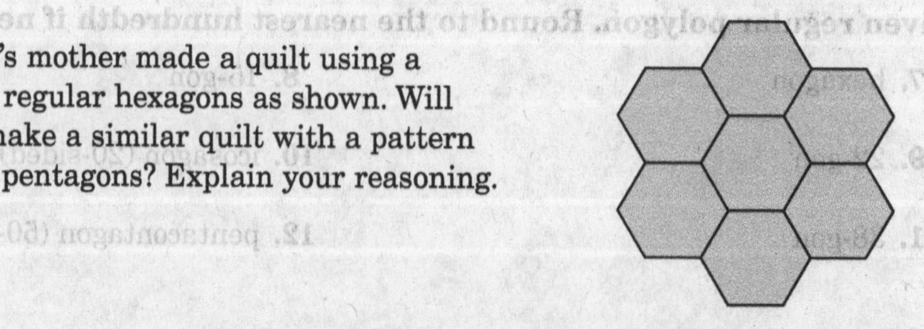

6-4 Study Guide and Intervention

Congruent Polygons

Two polygons are **congruent** if all pairs of corresponding angles are congruent and all pairs of corresponding sides are congruent. The letters identifying each polygon are written so that corresponding vertices appear in the same order.

Example 1 **Determine whether the triangles shown are congruent. If so, name the corresponding parts and write a congruence statement.**

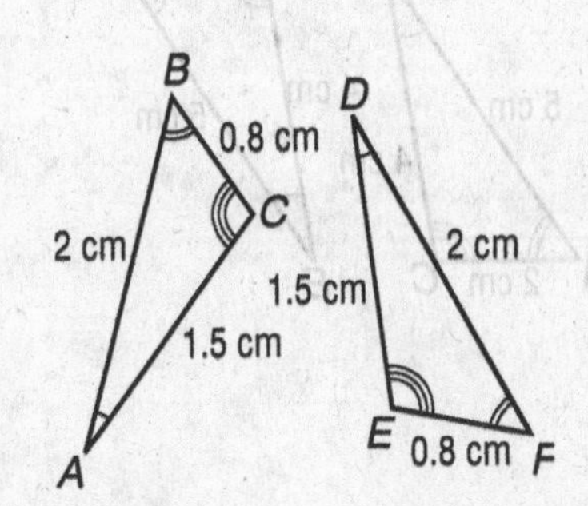

Angles The arcs indicate that $\angle A \cong \angle D$, $\angle B \cong \angle F$, and $\angle C \cong \angle E$.

Sides The side measures indicate that $\overline{AB} \cong \overline{DF}$, $\overline{BC} \cong \overline{FE}$, and $\overline{CA} \cong \overline{ED}$.

Since all pairs of corresponding sides and angles are congruent, the two triangles are congruent. One congruence statement is $\triangle ABC \cong \triangle DFE$.

Examples **In the figure, $\triangle MNB \cong \triangle LJH$.**

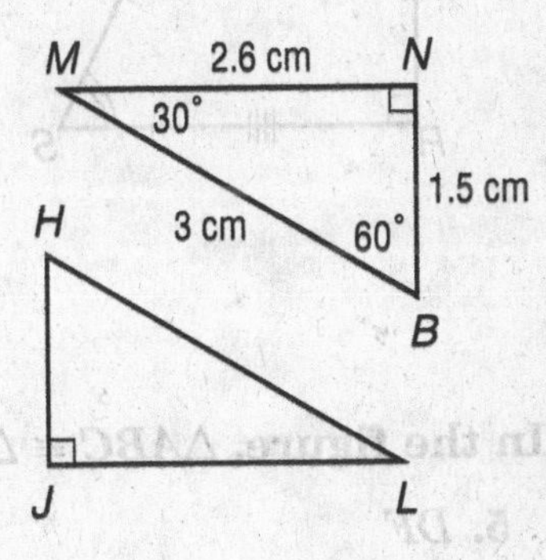

2 **Find *LJ*.**
$\overline{MN}$ corresponds to $\overline{LJ}$ So, $\overline{MN} \cong \overline{LJ}$.
Since $MN = 2.6$ centimeters, $LJ = 2.6$ centimeters.

3 **Find $m\angle H$.**
According to the congruence statement, $\angle B$ and $\angle H$ are corresponding angles. So, $\angle B \cong \angle H$. Since $m\angle B = 60°$, $m\angle H = 60°$.

Exercises

Determine whether the polygons shown are congruent. If so, name the corresponding parts and write a congruence statement.

1.

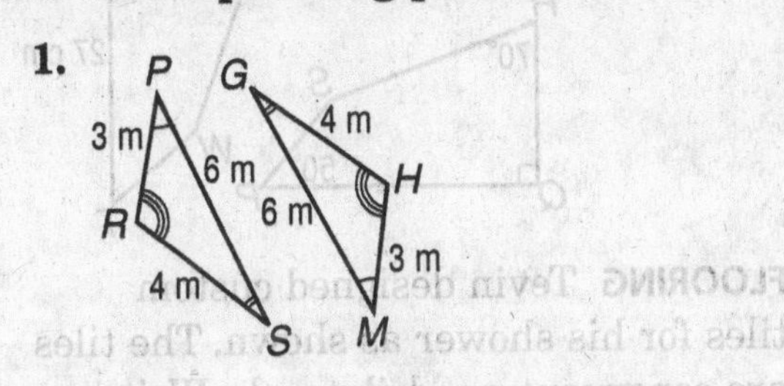

In the figure, $\triangle GFD \cong \triangle TRE$. Find each measure.

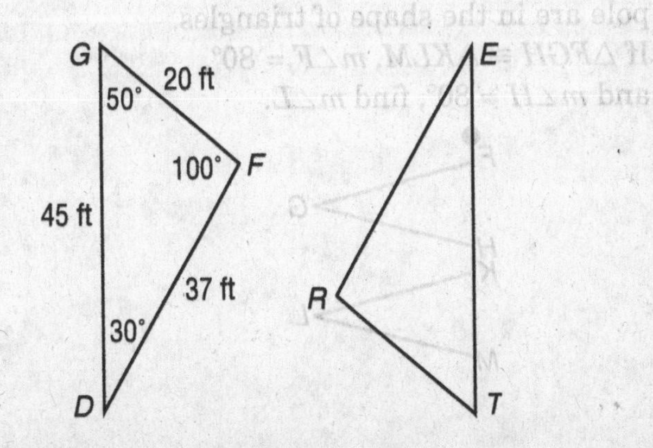

2. $m\angle R$

3. RT

4. $m\angle E$

6-4 Practice

Congruent Polygons

Determine whether the polygons are congruent. If so, name the corresponding parts and write a congruence statement.

1.

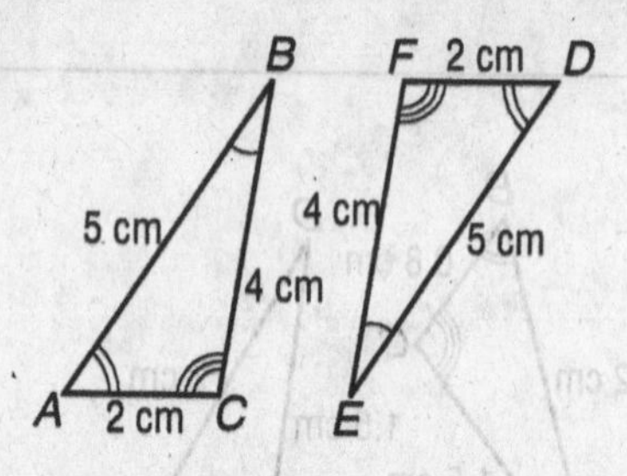

2.

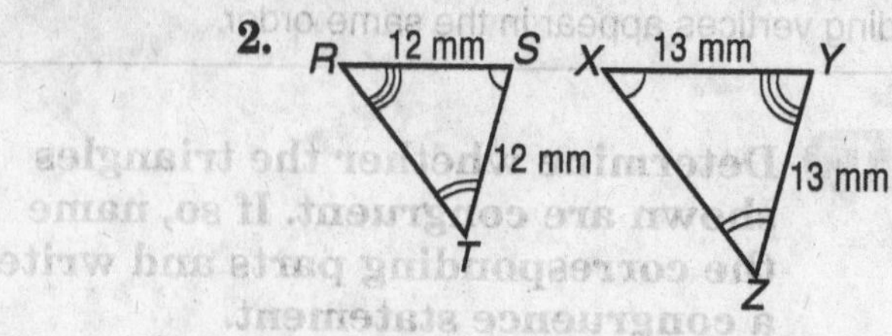

3.

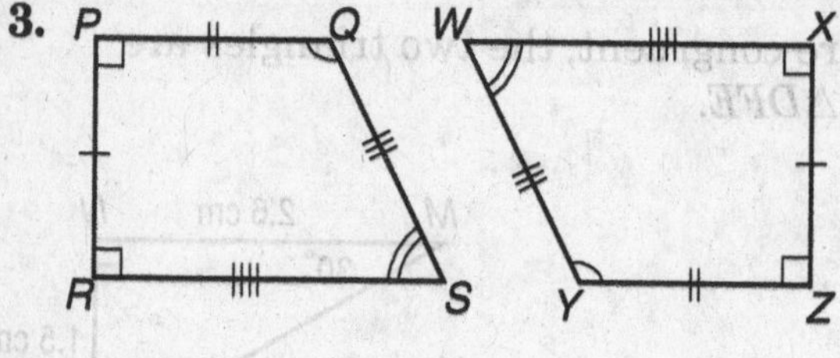

4.

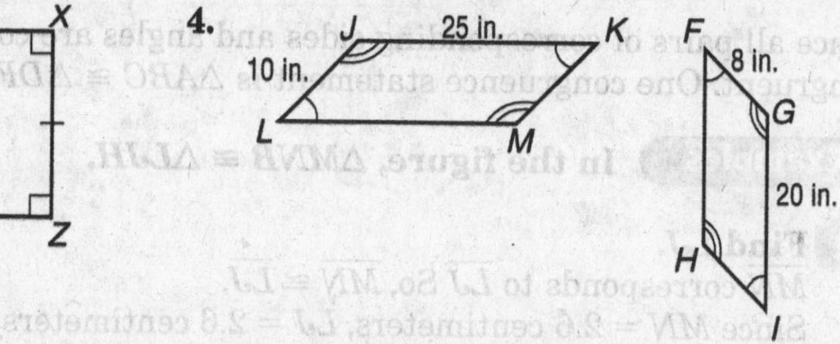

In the figure, $\triangle ABC \cong \triangle DEF$. Find each measure.

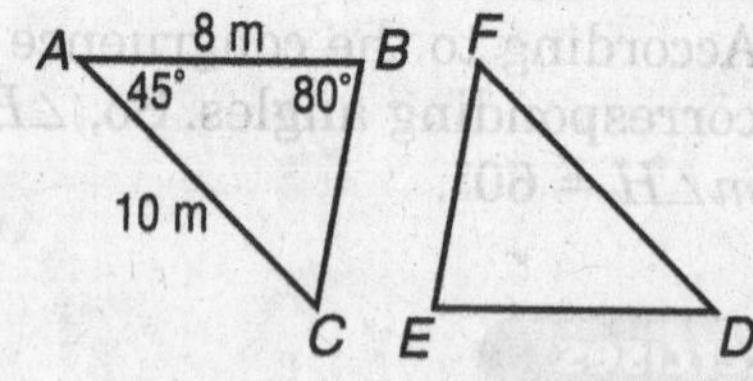

5. DF
6. DE
7. $m\angle D$
8. $m\angle E$

In the figure, quadrilateral $PQRS$ is congruent to quadrilateral $TUVW$. Find each measure.

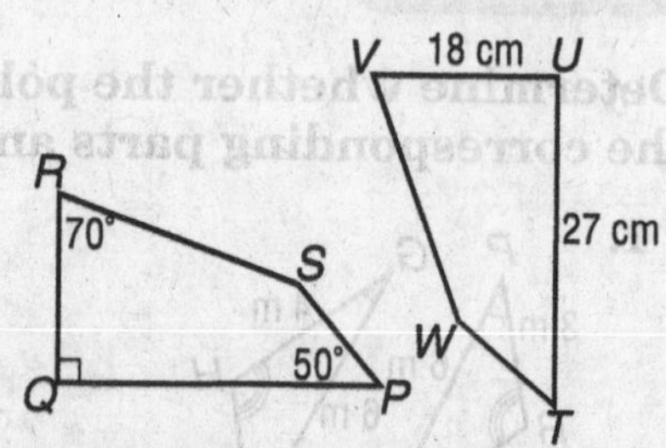

9. PQ
10. QR
11. $m\angle U$
12. $m\angle V$

13. **FLAGS** The two flags flying on the pole are in the shape of triangles. If $\triangle FGH \cong \triangle KLM$, $m\angle F = 80°$, and $m\angle H = 80°$, find $m\angle L$.

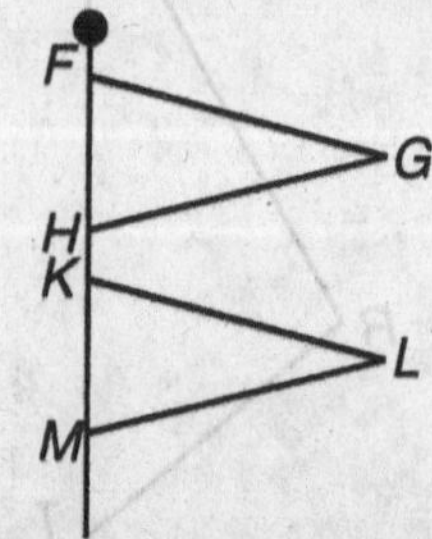

14. **FLOORING** Tevin designed custom tiles for his shower as shown. The tiles are congruent quadrilaterals. Write a congruence statement. Then find $m\angle J$ if $m\angle A = 90°$, $m\angle B = 60°$, and $m\angle D = 90°$.

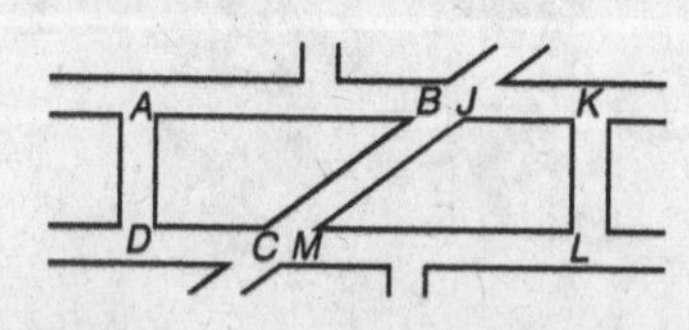

6-5 Study Guide and Intervention

Symmetry

A figure has **line symmetry** if it can be folded over a line so that one half of the figure matches the other half. This fold line is called the **line of symmetry**. Some figures have more than one line of symmetry.

Example 1 **Determine whether the figure has line symmetry. If it does, trace the figure and draw all lines of symmetry. If not, write *none*.**

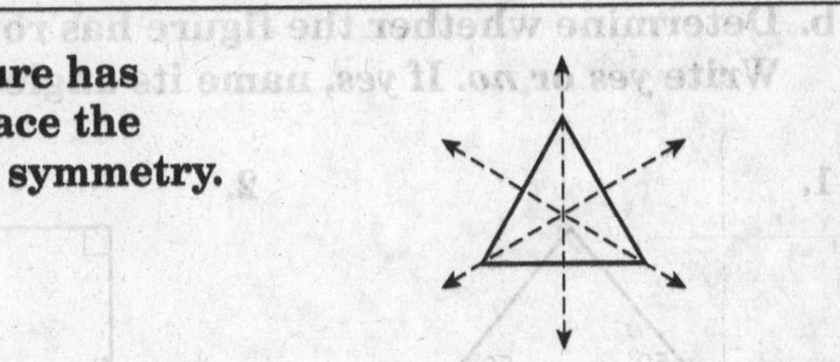

This figure has three lines of symmetry.

A figure has **rotational symmetry** if it can be rotated or turned less than 360° about its center so that the figure looks exactly as it does in its original position. The degree measure of the angle through which the figure is rotated is called the **angle of rotation**.

Example 2 **Determine whether the figure has rotational symmetry. Write *yes* or *no*. If *yes*, name its angles of rotation.**

Yes, this figure has rotational symmetry. It matches itself after being rotated 180°.

Exercises

For Exercises 1–6, complete parts a and b for each figure.

a. Determine whether the figure has line symmetry. If it does, draw all lines of symmetry. If not, write *none*.

b. Determine whether the figure has rotational symmetry. Write *yes* or *no*. If *yes*, name its angles of rotation.

1.

2.

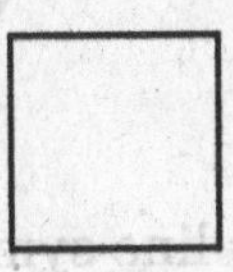

3.

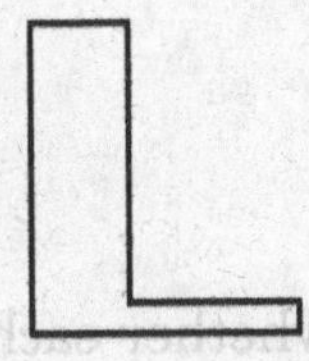

4.

5.

6.

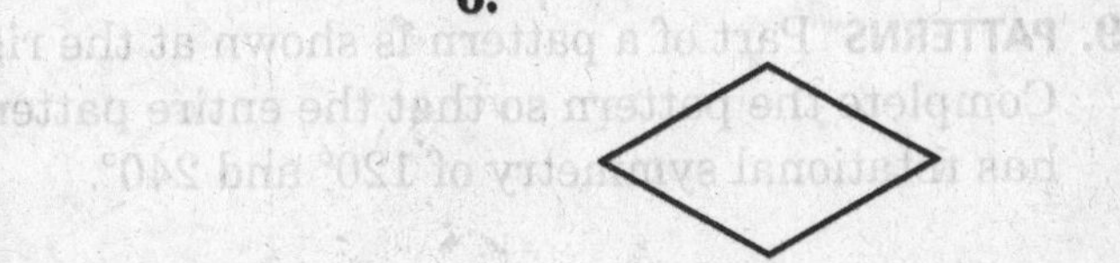

Lesson 6-5

NAME ______________________ DATE ____________ PERIOD _____

6-5 Practice

Symmetry

GEOMETRY For Exercises 1–6, complete parts a and b for each figure.

a. Determine whether the figure has line symmetry. If it does, draw all lines of symmetry. If not, write *none*.

b. Determine whether the figure has rotational symmetry. Write *yes* or *no*. If *yes*, name its angle(s) of rotation.

1. **2.** **3.**

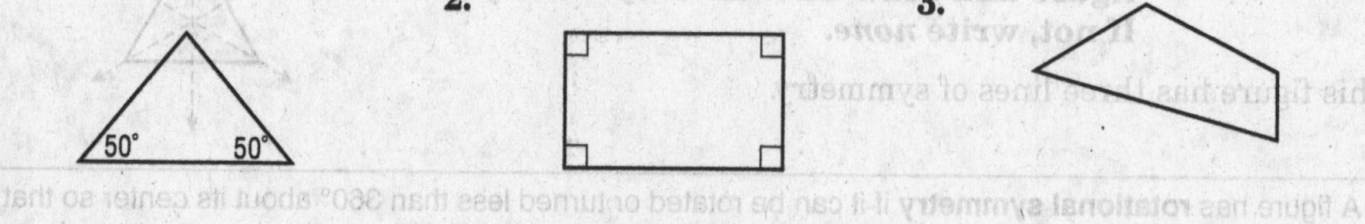

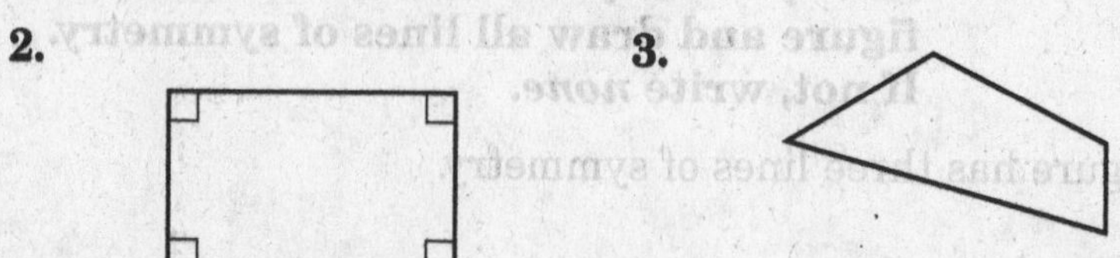

4. **5.** **6.**

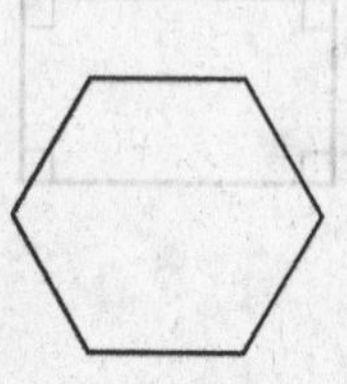

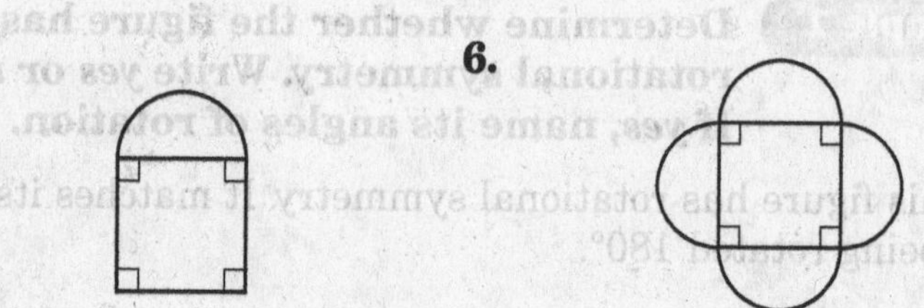

GREEK LETTERS For Exercises 7 and 8, use the Greek letters below.

a. gamma **b.** chi **c.** psi **d.** omega

Γ   

7. Determine whether each Greek letter has line symmetry. If it does, draw all lines of symmetry. If not, write *none*.

8. Which of the Greek letters could be rotated and still look the same? If any, name the angle(s) of rotation.

9. PATTERNS Part of a pattern is shown at the right. Complete the pattern so that the entire pattern has rotational symmetry of 120° and 240°.

6-6 Study Guide and Intervention

Reflections

When a figure is reflected across a line, every point on the **reflection** is the same distance from the line of reflection as the corresponding point on the original figure. The image is congruent to the original figure, but the orientation of the image is different from that of the original figure.

Example 1 **Draw the image of quadrilateral *ABCD* after a reflection over the given line.**

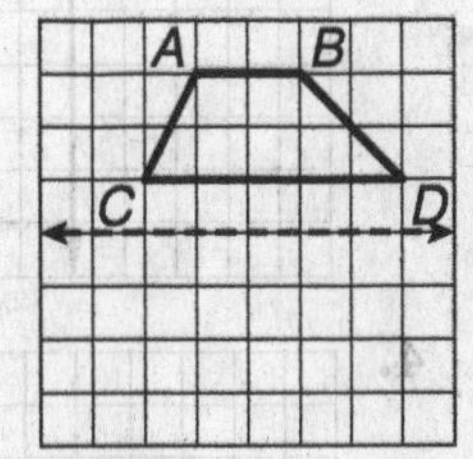

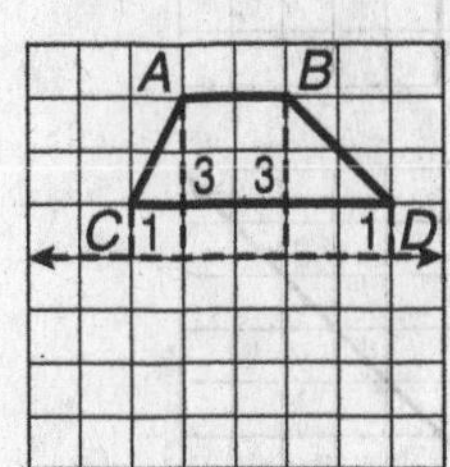

Step 1 Count the number of units between each vertex and the line of reflection.

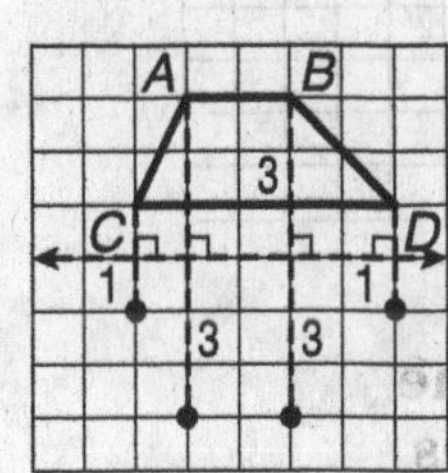

Step 2 To find the corresponding point for vertex *A*, move along the line through vertex *A* perpendicular to the line of reflection until you are 3 units from the line on the opposite side. Draw a point and label it *A*′. Repeat for each vertex.

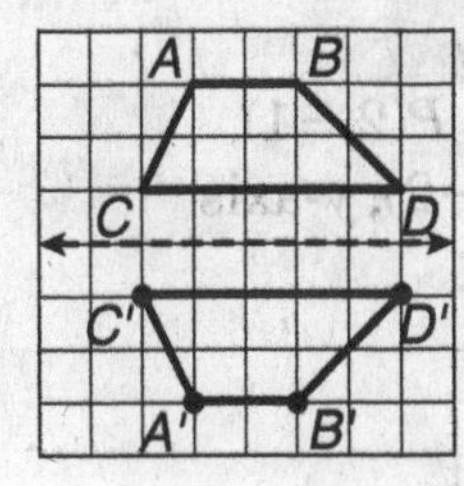

Step 3 Connect the new vertices to form quadrilateral *A*′*B*′*C*′*D*′.

Notice that if you move along quadrilateral *ABCD* from *A* to *B* to *C* to *D*, you are moving in the clockwise direction. However, if you move along quadrilateral *A*′*B*′*C*′*D*′ from *A*′ to *B*′ to *C*′ to *D*′, you are moving in the counterclockwise direction. A figure and its reflection have opposite orientations.

Exercises

Draw the image of the figure after a reflection over the given line.

1. 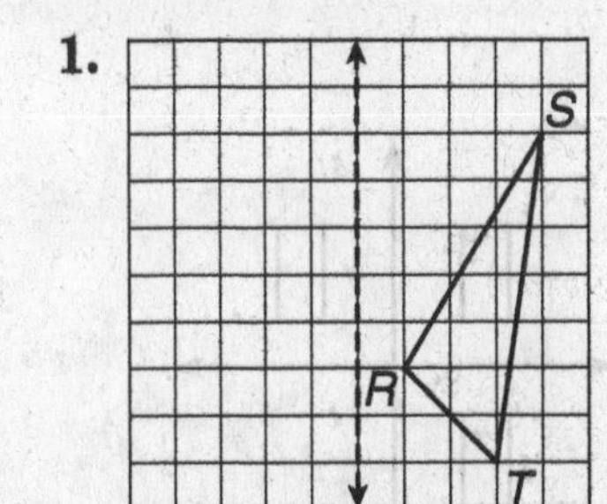

2.
G
F
I
H

3. 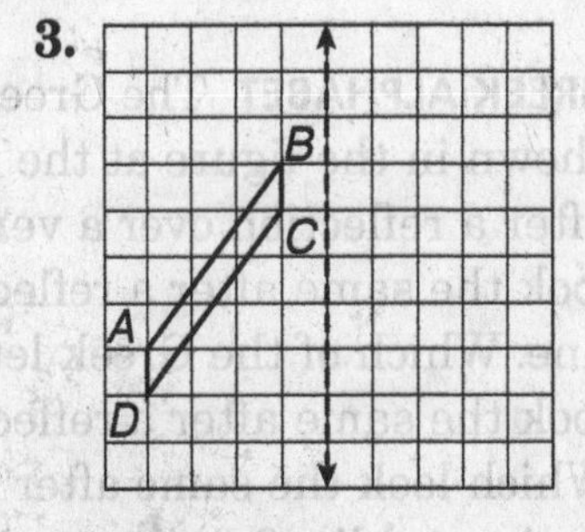

6-6 Practice

Reflections

Draw the image of the figure after a reflection across the given line.

1\.

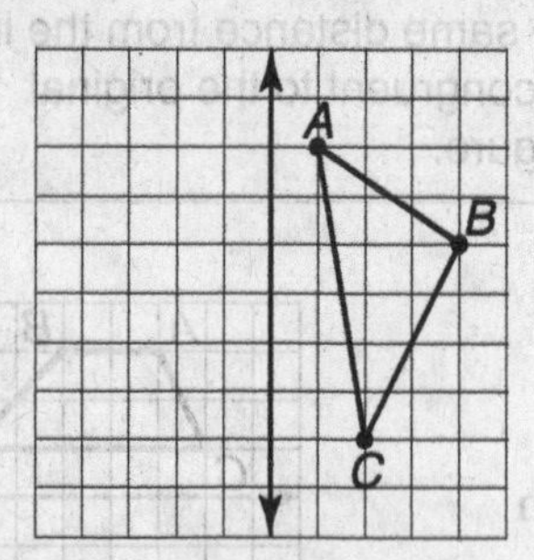

2\.

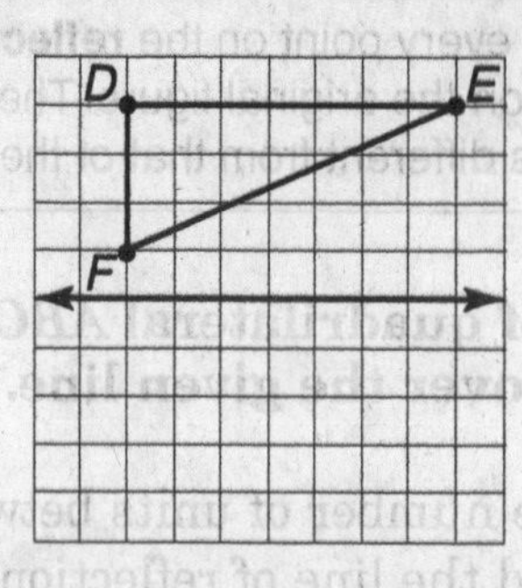

3\.

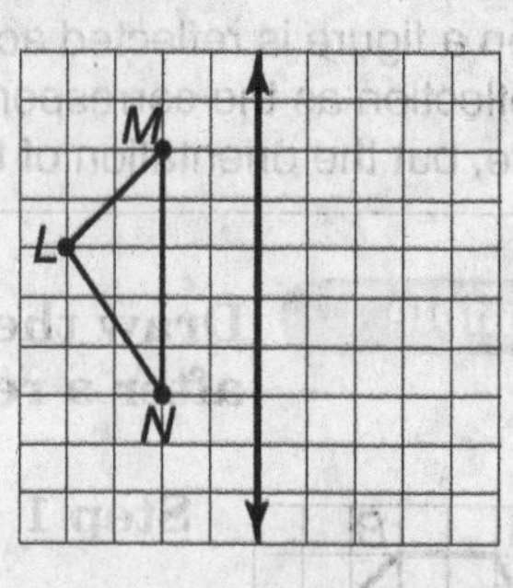

4\.

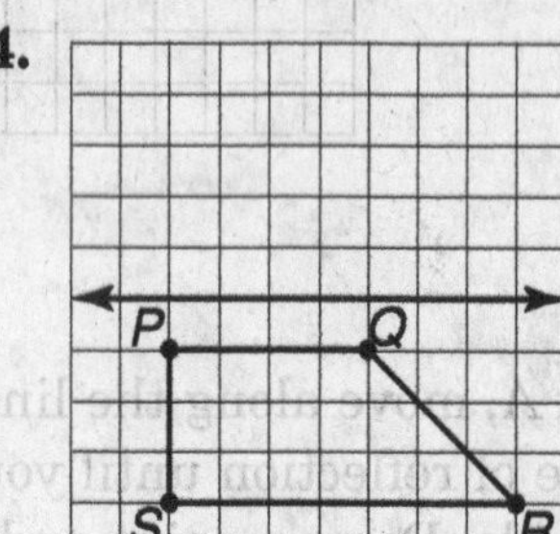

5\.

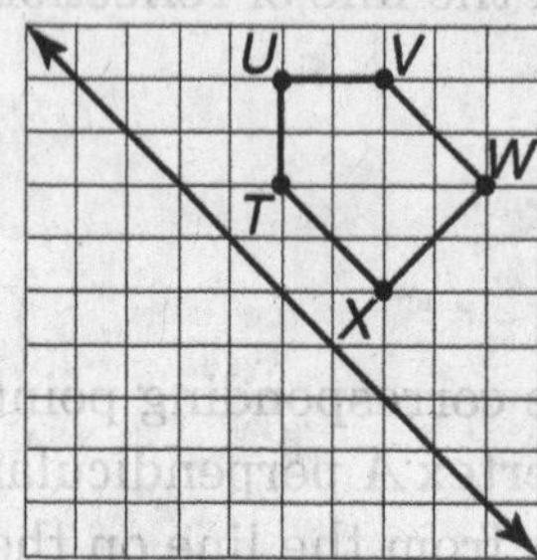

6\.

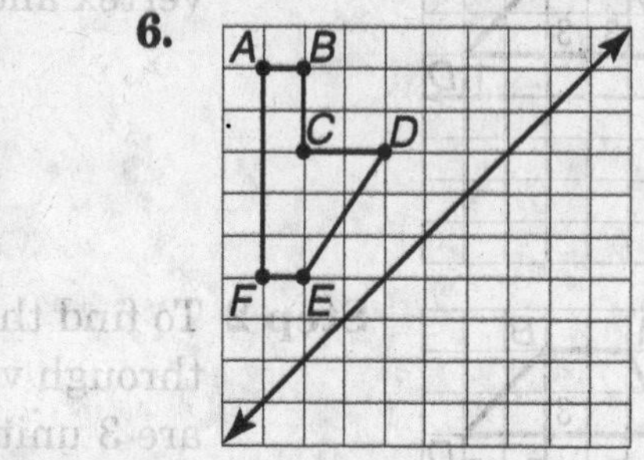

Graph the figure with the given vertices. Then graph the image of the figure after a refection over the given axis, and write the coordinates of the image's vertices.

7\. triangle FGH with vertices $F(-1, 1)$, $G(-3, 3)$, and $H(-4, 2)$; y-axis

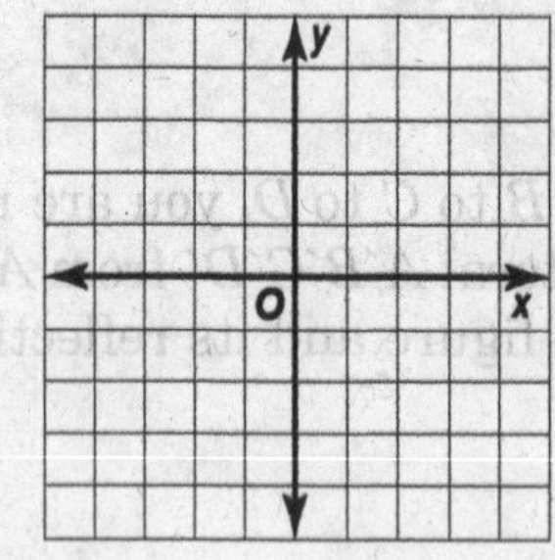

8\. square $PQRS$ with vertices $P(2, -1)$, $Q(3, -2)$, $R(2, -3)$, and $S(1, -2)$; y-axis

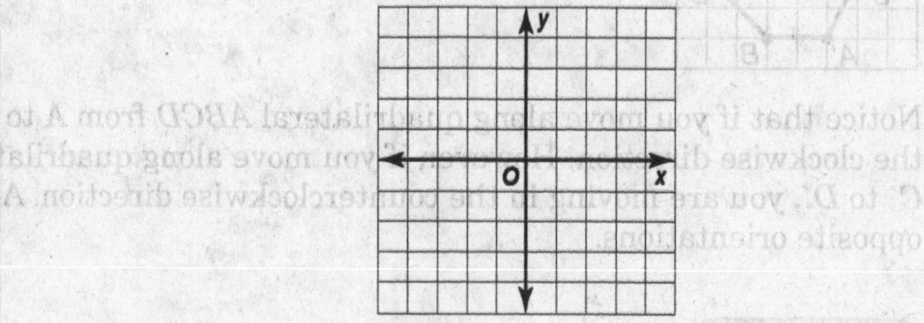

9\. **GREEK ALPHABET** The Greek letter pi, Π, shown in the figure at the right, looks the same after a reflection over a vertical line. It does not look the same after a reflection over a horizontal line. Which of the Greek letters Γ, Φ, Θ, Z, Ω, and Ψ look the same after a reflection over a vertical line? Which look the same after a reflection over a horizontal line?

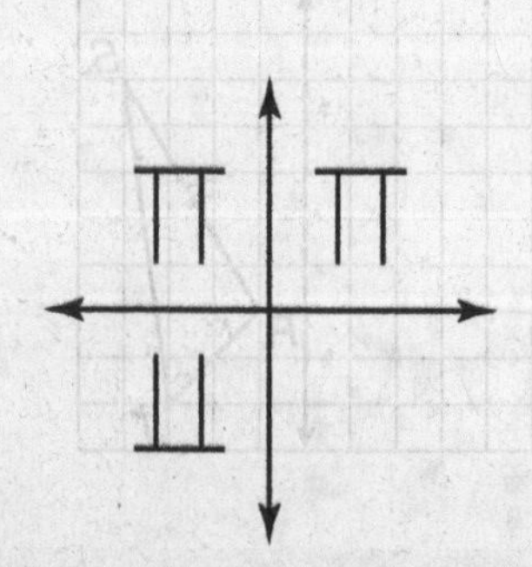

6-7 Study Guide and Intervention

Translations

When a figure is translated, every point is moved the same distance and in the same direction. The translated figure is congruent to the original figure and has the same orientation.

Example **Draw the image of quadrilateral *ABCD* after a translation 2 units right and 3 units up.**

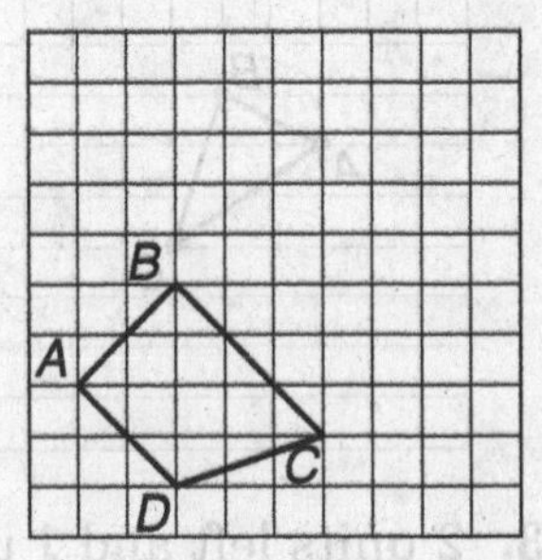

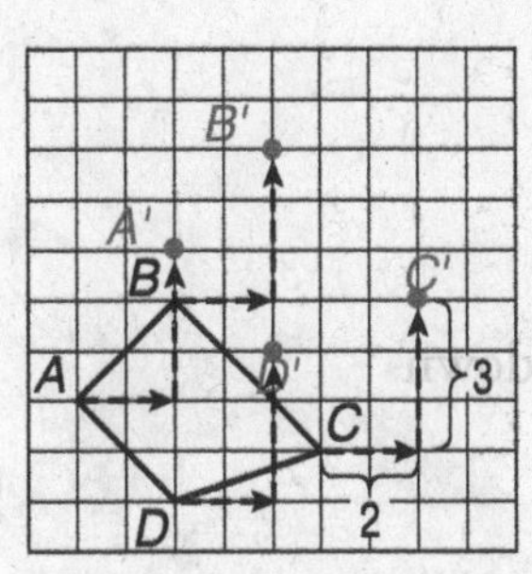

Step 1 To find the corresponding point for vertex *A*, start at *A* and move 2 units to the right along the horizontal grid line and then move up 3 units along the vertical grid line. Draw a point and label it *A´*. Repeat for each vertex.

Step 2 Connect the new vertices to form quadrilateral *A´B´C´D´*.

Exercises

Draw the image of the figure after the indicated translation.

1. 5 units right and 4 units down

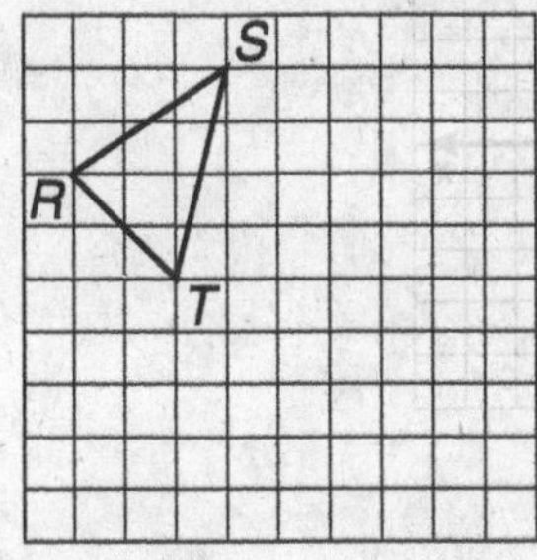

2. 3 units left and 2 units up

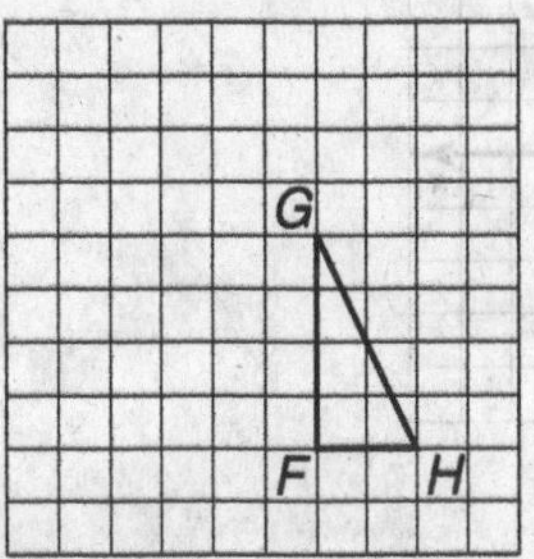

3. 2 units left and 3 units down

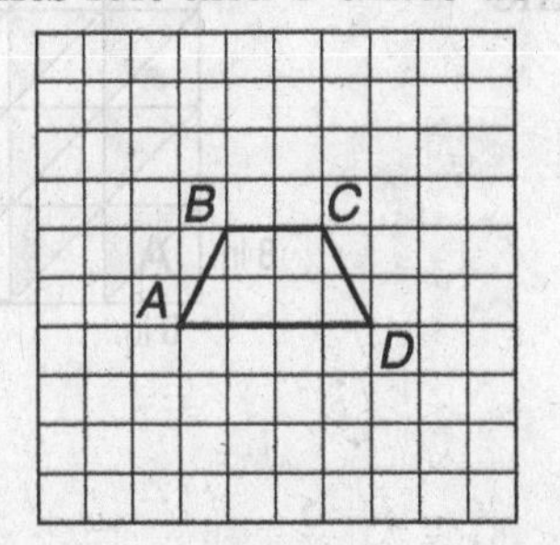

4. 2 units right and 1 unit up

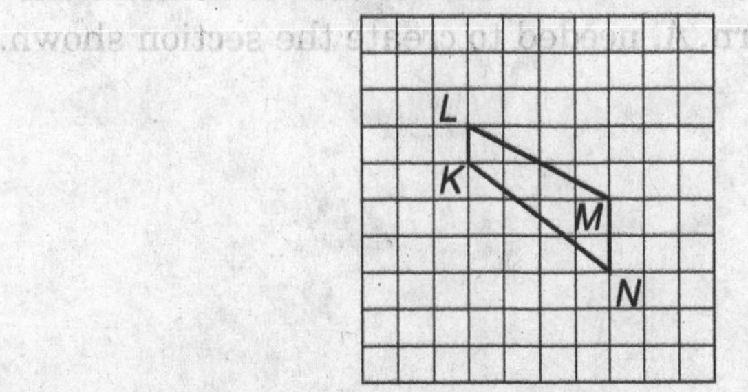

NAME ______________________________ DATE ____________ PERIOD _____

6-7 Practice

Translations

Draw the image of the figure after the indicated translation.

1. 3 units right and 2 units up

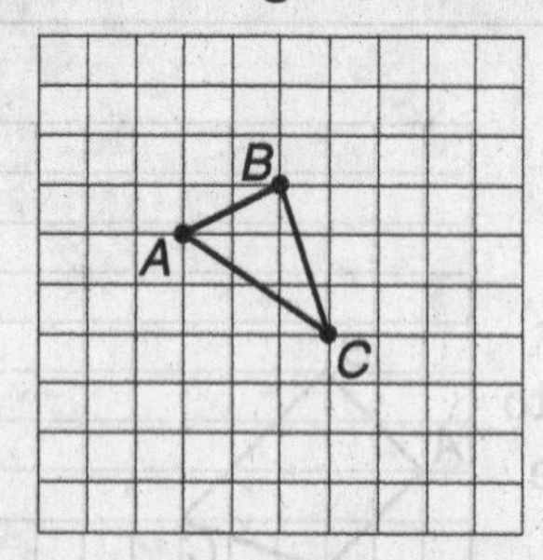

2. 5 units right and 3 units down

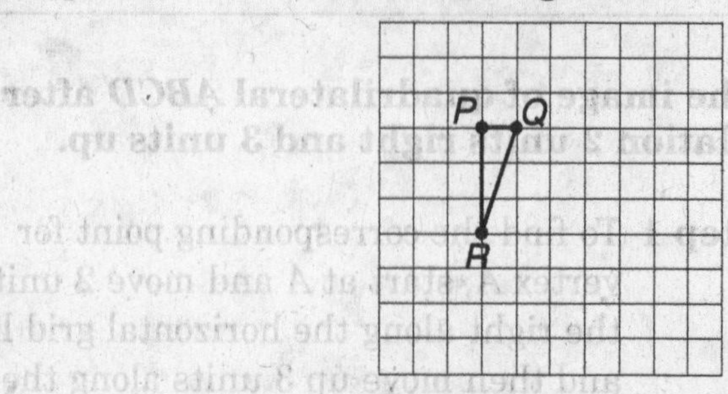

3. 2 units left and 1 unit up

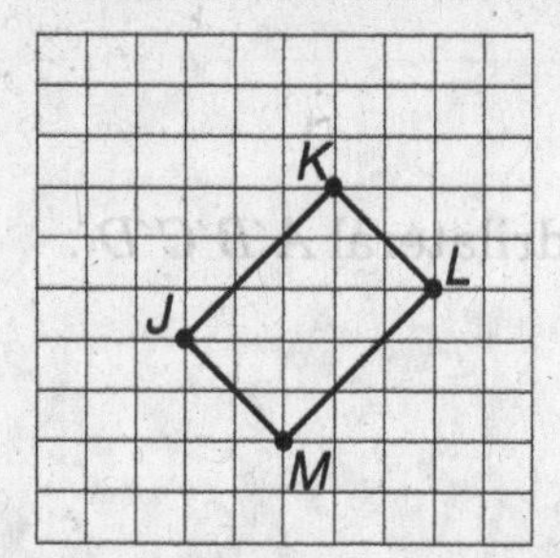

4. 4 units left and 2 units down

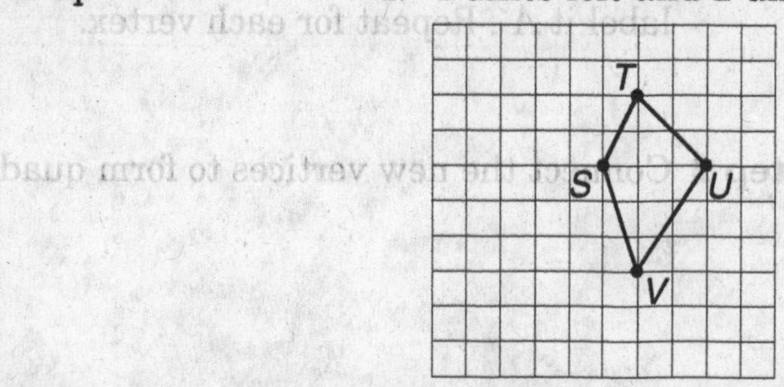

Graph the figure with the given vertices. Then graph the image of the figure after the indicated translation, and write the coordinates of its vertices.

5. $\triangle FGH$ with vertices $F(1, 3)$, $G(2, 4)$, and $H(3, 2)$; translated 3 units left and 1 unit down

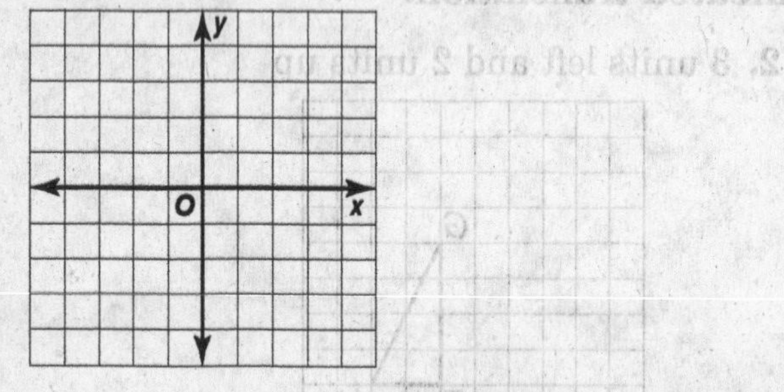

6. rectangle $PQRS$ with vertices $P(-4,-1)$, $Q(0, 1)$, $R(1,-1)$, and $S(-3,-3)$ translated 2 units right and 3 units up

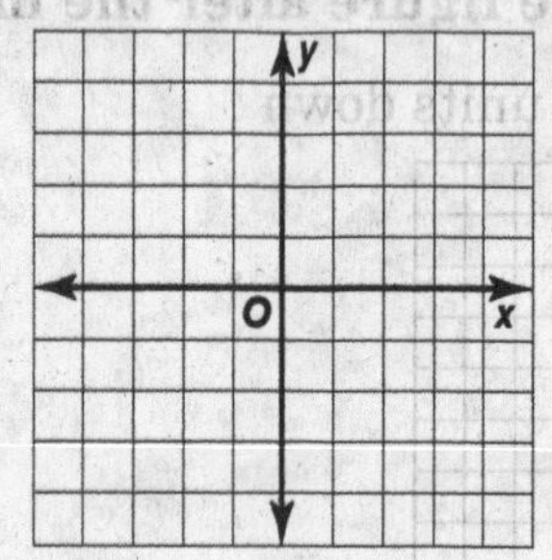

7. QUILTS The quilt design at the right is a traditional American design. Describe the minimum number of translations of the original pattern, A, needed to create the section shown.

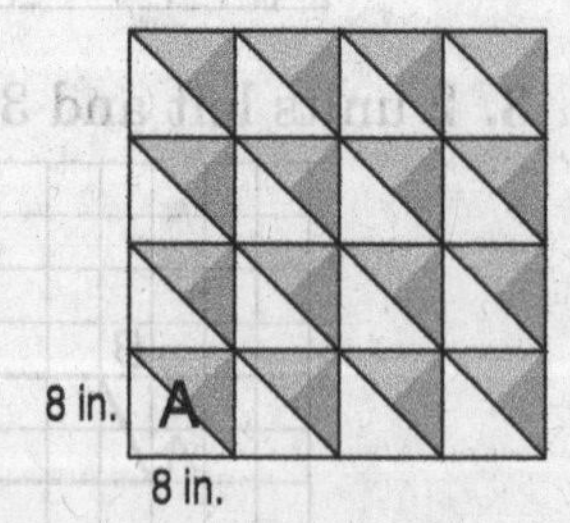

7-1 Study Guide and Intervention

Circumference and Area of Circles

The **circumference** C of a circle is equal to its diameter d times π or 2 times the radius r times π, or $C = \pi d$ or $C = 2\pi r$.

The **area** A of a circle is equal to π times the square of the radius r, or $A = \pi r^2$.

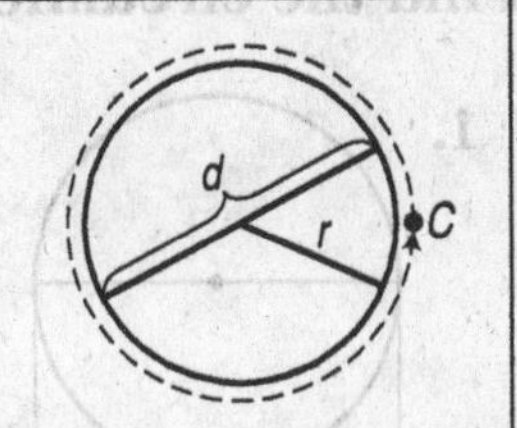

Examples **Find the circumference of each circle. Use 3.14 for π. Round to the nearest tenth.**

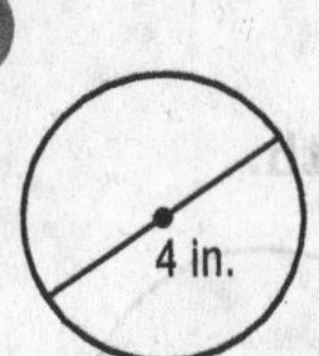

$C = \pi d$	Circumference of a circle
$C = \pi \cdot 4$	Replace d with 4.
$C = 4\pi$	This is the exact circumference.
$C \approx 4 \cdot 3.14$ or 12.6	Replace π with 3.14 and multiply.

The circumference is about 12.6 inches.

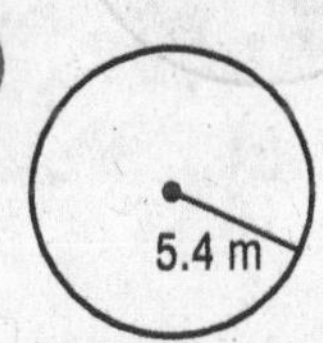

$C = 2\pi r$	Circumference of a circle
$C \approx 2 \cdot 3.14 \cdot 5.4$	Replace r with 5.4.
$C \approx 33.9$	Replace π with 3.14 and multiply.

The circumference is about 33.9 meters.

Example 3 **Find the area of the circle. Use 3.14 for π. Round to the nearest tenth.**

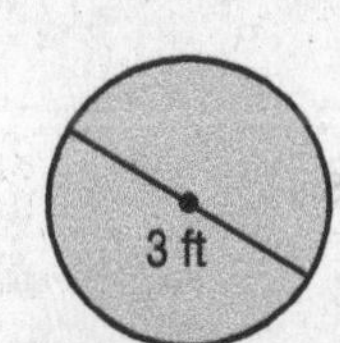

$A = \pi r^2$	Area of a circle
$A \approx 3.14(1.5)^2$	Replace π with 3.14 and r with half of 3 or 1.5.
$A \approx 3.14 \cdot 2.25$	Evaluate $(1.5)^2$.
$A \approx 7.1$	Multiply.

The area is about 7.1 square feet.

Exercises

Find the circumference and area of each circle. Use 3.14 for π. Round to the nearest tenth.

1. 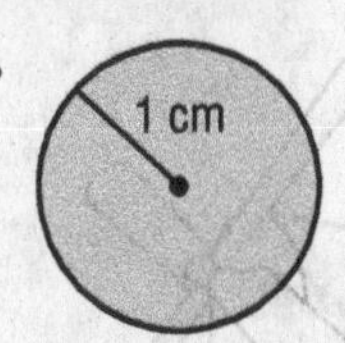

2. 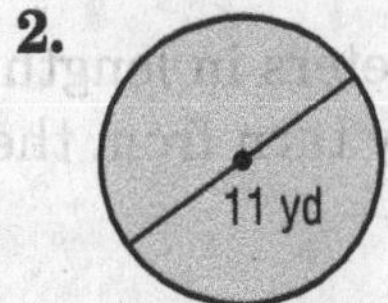

3. 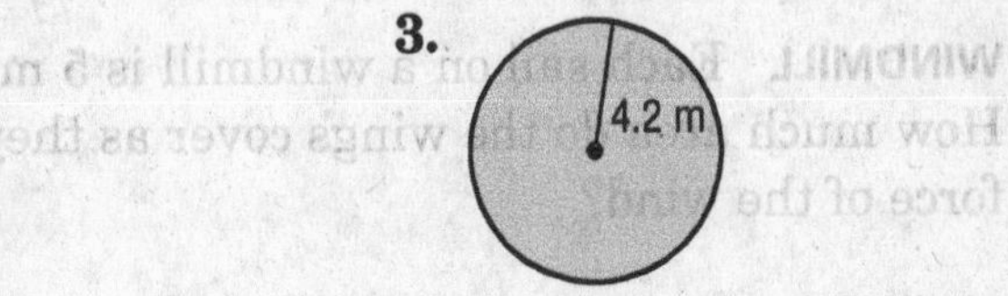

4. The diameter is 9.3 meters.

5. The radius is 6.9 millimeters.

6. The diameter is 15.7 inches.

NAME ______________________________ DATE ____________ PERIOD _____

7-1 Practice

Circumference and Area of Circles

Find the circumference of each circle. Use 3.14 for π. Round to the nearest tenth.

1.

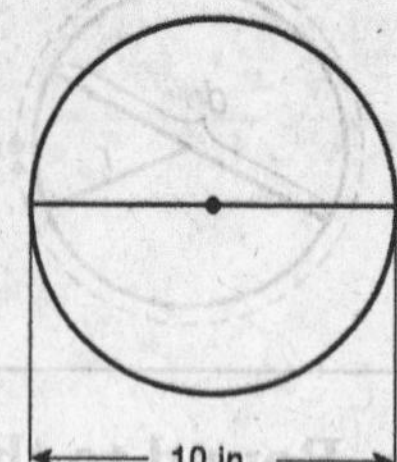

2.

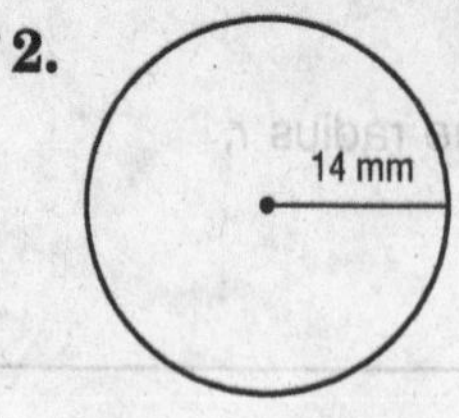

3.

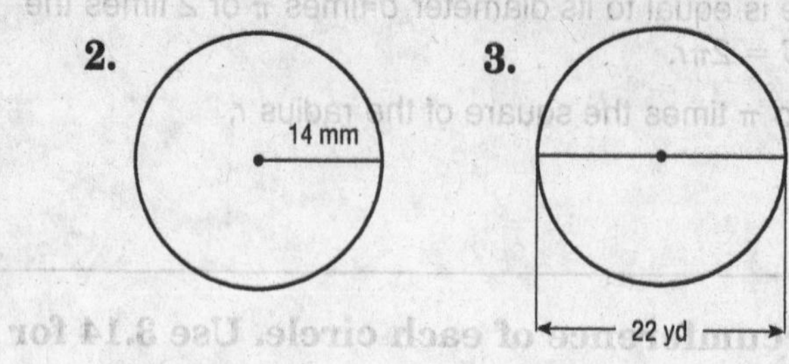

4.

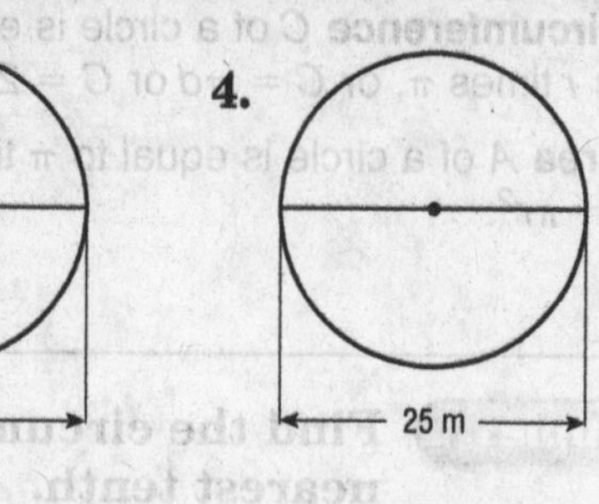

Find the area of each circle. Use 3.14 for π. Round to the nearest tenth.

5.

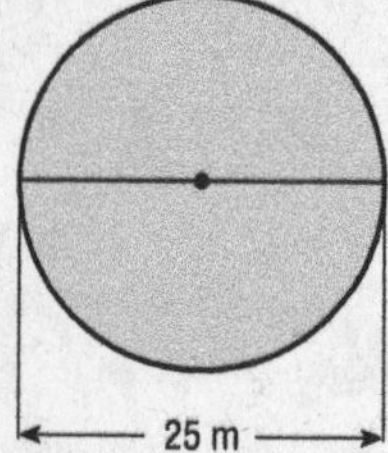

6.

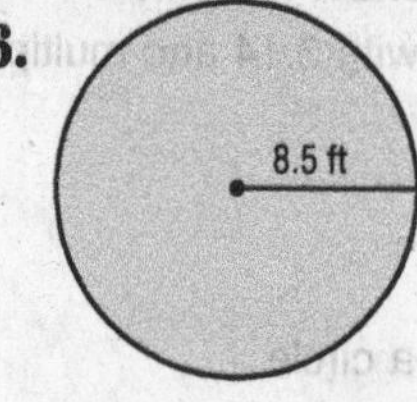

7.

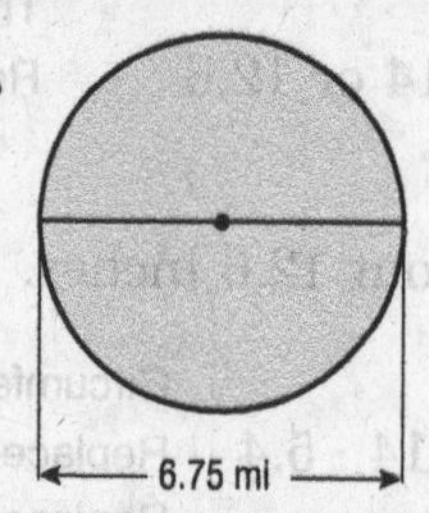

8.

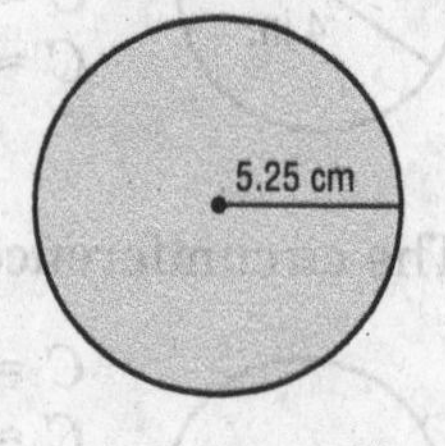

Find the circumference and area of each circle. Round to the nearest tenth.

9. The diameter is 8 centimeters.

10. The radius is 4.7 inches.

11. The radius is 0.9 feet.

12. The diameter is 6.8 kilometers.

Another approximate value for π is $\frac{22}{7}$. Use this value to find the circumference and area of each circle.

13. The diameter is 14 yards.

14. The radius is $1\frac{1}{6}$ millimeters.

15. WINDMILL Each sail on a windmill is 5 meters in length. How much area do the wings cover as they turn from the force of the wind?

16. ALGEBRA Find the radius of a circle if its area is 314 square miles.

7-2 Study Guide and Intervention

Problem-Solving Investigation: Solve a Simpler Problem

Example 1 **Gift cards come in packages of 12 and envelopes come in packages of 15. Meagan needs to send 600 cards in envelopes. How many packages of each kind should she buy?**

Understand Meagan needs the same number of cards and envelopes.

Plan Find out how many packages are needed for 300 cards in envelopes.

Solve $12c = 300$ $15e = 300$

$c = 25$ $e = 20$

Multiply the answers by 2.

Check $2 \times 25 = 50$ packages of cards $2 \times 20 = 40$ packages of envelopes

Meagan should buy 50 packages of cards and 40 packages of envelopes.

Example 2 **How many triangles of any size are in the figure at the right?**

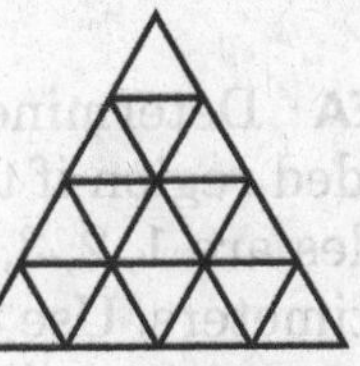

Understand We need to find how many triangles are in the figure.

Plan Draw a simpler diagram.

Solve

9 Count the smallest triangles, which have 1 triangle per side.
3 Count the next largest triangles, which have 2 triangles per side.
1 Count the largest triangle, which has 3 triangles per side.
13 Add together to find the total triangles of any size.

Check Now repeat the steps for the original problem.

16 Count the smallest triangles, which have 1 triangle per side.
7 Count the next largest triangles, which have 2 triangles per side.
3 Count the next largest triangles, which have 3 triangles per side.
1 Count the largest triangle, which has 4 triangles per side.
27 Add together to find the total triangles of any size.

Exercises **For Exercises 1–3, solve a simpler problem.**

1. Hot dogs come in packages of 10 and buns come in packages of 8. How many packages of each will Mindy need to provide 640 hot dogs for a street fair?

2. Mark can plant 3 tree saplings in an hour and Randy can plant 5 tree saplings in an hour. Working together, how long will it take them to plant 80 tree saplings?

3. A restaurant has 18 square tables that can be pushed together to form one long table for large parties. Each square table can seat 2 people per side. How many people can be seated at the combined tables?

Lesson 7-2

NAME ________________________________ DATE ____________ PERIOD _____

7-2 Practice

Problem-Solving Investigation: Solve a Simpler Problem

Mixed Problem Solving

Use the solve a simpler problem strategy to solve Exercises 1 and 2.

1. **ASSEMBLY** A computer company has two locations that assemble computers. One location assembles 13 computers in an hour and the other location assembles 12 computers in an hour. Working together, how long will it take both locations to assemble 80 computers?

2. **AREA** Determine the area of the shaded region if the radii of the six circles are 1, 2, 3, 4, 5, and 10 centimeters. Use 3.14 for π. Round to the nearest tenth if necessary.

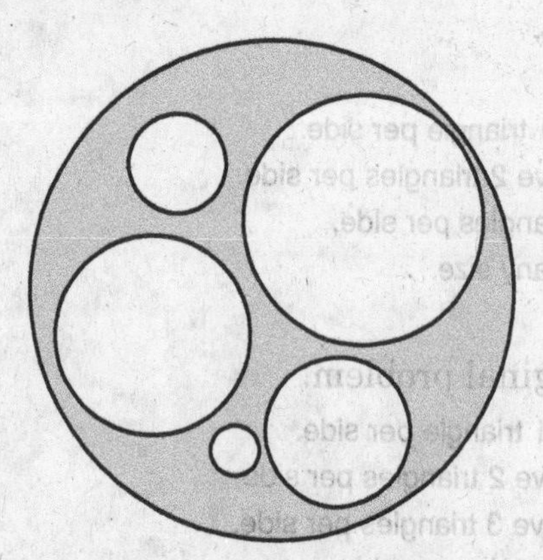

Use any strategy to solve Exercises 3–6. Some strategies are shown below.

PROBLEM-SOLVING STRATEGIES
• Look for a pattern.
• Use a Venn diagram.
• Solve a simpler problem.

3. **NUMBER SENSE** Find the sum of all the even numbers from 2 to 50, inclusive.

4. **ANALYZE TABLES** Mr. Brown has $1,050 to spend on computer equipment. Does Mr. Brown have enough money to buy the computer, scanner, and software if a 20% discount is given and the sales tax is 5%? Explain.

Item	Cost
Computer	$899
Scanner	$54
Software	$278

5. **COPIER** The counter on a business copier read 18,678 at the beginning of the week and read 20,438 at the end of the week. If the business was in operation 40 hours that week, what was the average number of copies made each hour?

6. **HUMMINGBIRD** In normal flight a hummingbird can flap its wings 75 times each second. At this rate, how many times does a hummingbird flap it wings in a 20-minute flight?

7-3 Study Guide and Intervention

Area of Composite Figures

To find the area of a composite figure, separate the figure into shapes whose areas you know how to find. Then find the sum of these areas.

Example **Find the area of the composite figure.**

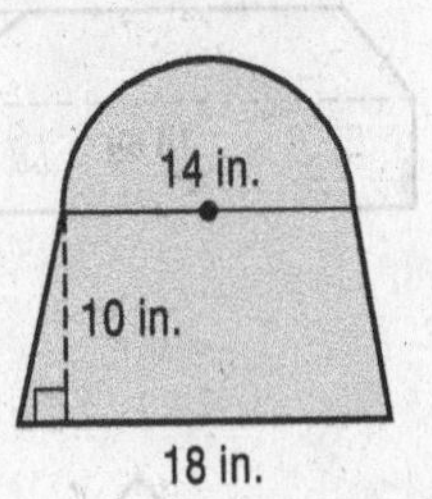

The figure can be separated into a semicircle and trapezoid.

Area of semicircle

$A = \frac{1}{2}\pi r^2$

$A = \frac{1}{2} \cdot 3.14 \cdot (7)^2$

$A \approx 77.0$

Area of trapezoid

$A = \frac{1}{2}h(b_1 + b_2)$

$A = \frac{1}{2} \cdot 10 \cdot (14 + 18)$

$A = 160$

The area of the figure is about 77.0 + 160 or 237 square inches.

Exercises

Find the area of each figure. Use 3.14 for π. Round to the nearest tenth if necessary.

1. 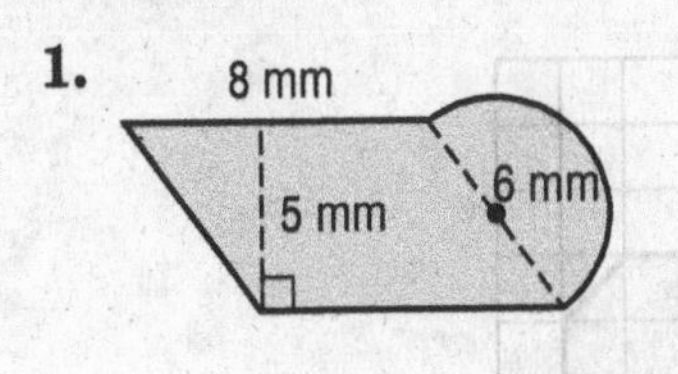

2. 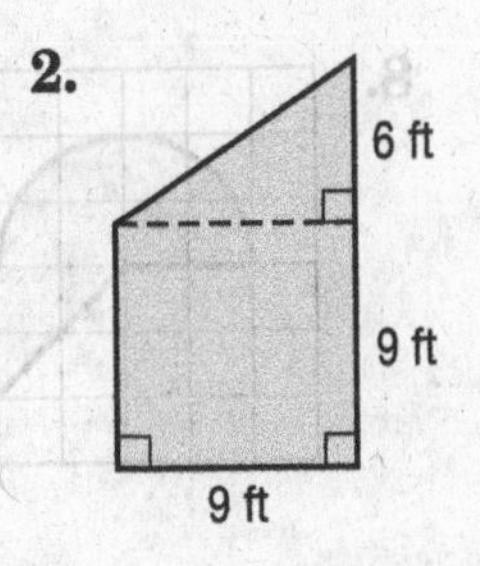

3. 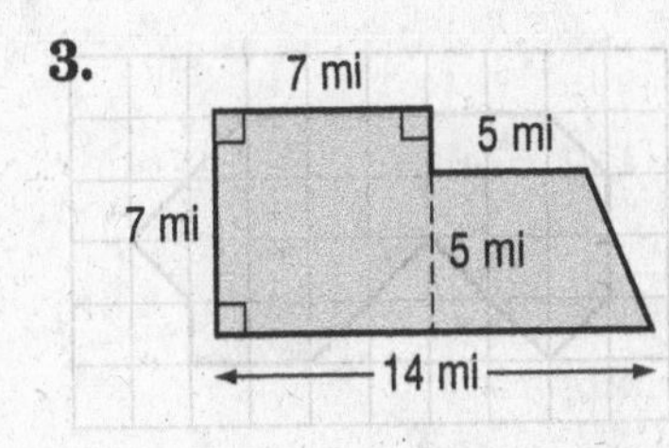

4. What is the area of a figure formed using a triangle with a base of 6 meters and a height of 11 meters and a parallelogram with a base of 6 meters and a height of 11 meters?

5. What is the area of a figure formed using a semicircle with a diameter of 8 yards and a square with sides of a length of 6 yards?

6. What is the area of a figure formed using a rectangle with a length of 9 inches and a width of 3 inches and a triangle with a base of 4 inches and a height of 13 inches?

7-3 Practice

Area of Composite Figures

Find the area of each figure. Use 3.14 for π. Round to the nearest tenth if necessary.

1.

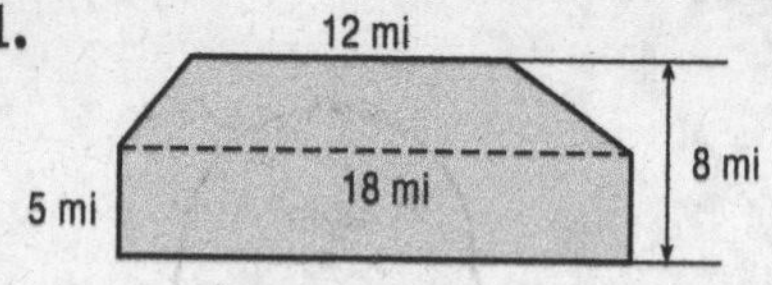

2.

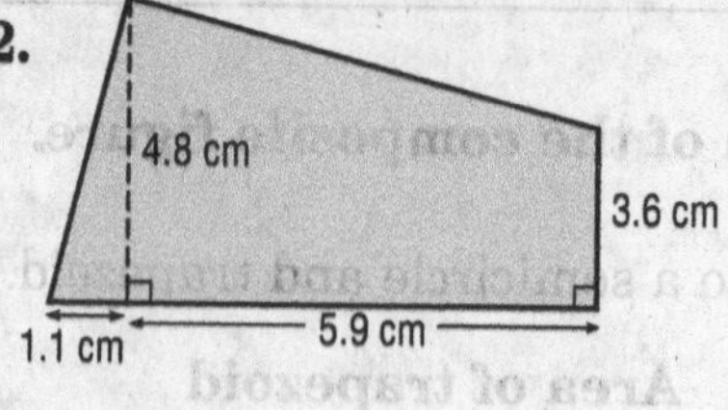

3.

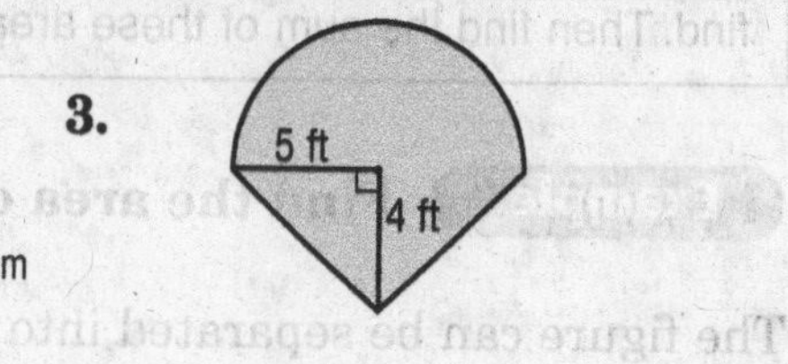

4.

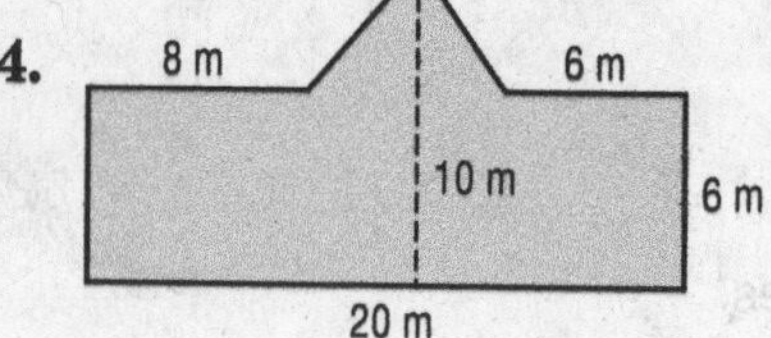

5.

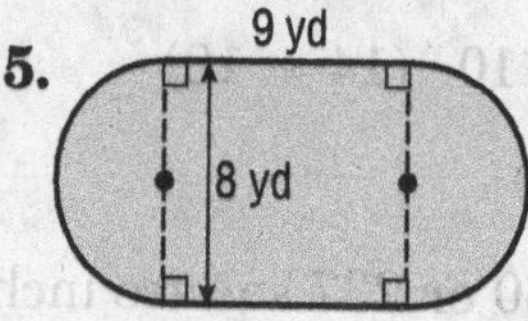

6.

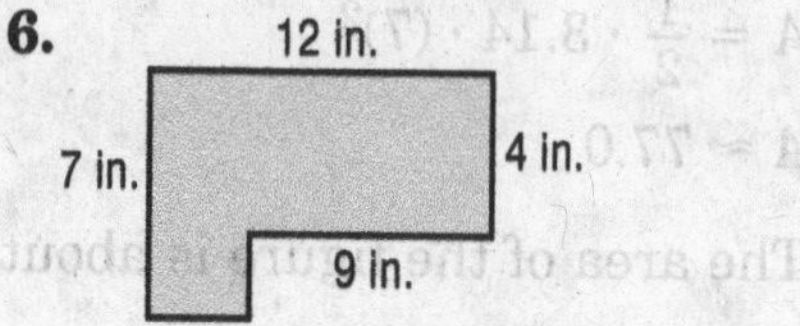

In each diagram, one square unit represents 10 square centimeters. Find the area of each figure. Round to the nearest tenth if necessary.

7.

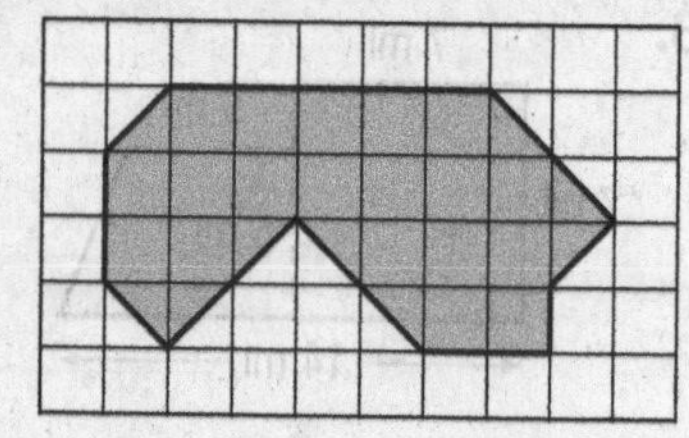

8.

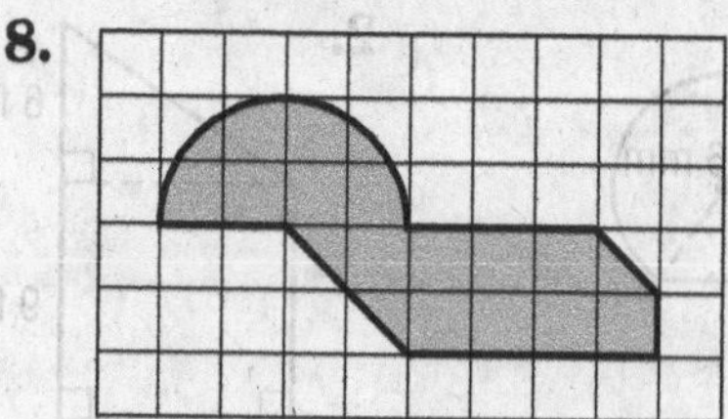

9. GAZEBO The Parks and Recreation department is building a gazebo in the local park with the dimensions shown in the figure. What is the area of the floor?

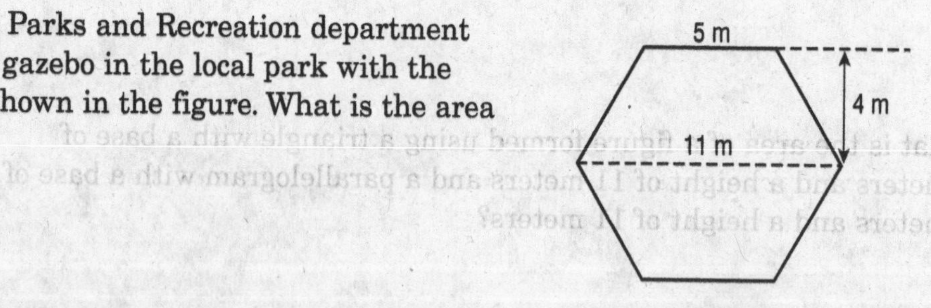

10. DECK The Pueyo family wants to paint the deck around their swimming pool with the dimensions shown in the figure. If a gallon covers 200 square feet, how many gallons of paint are needed to apply two coats of paint?

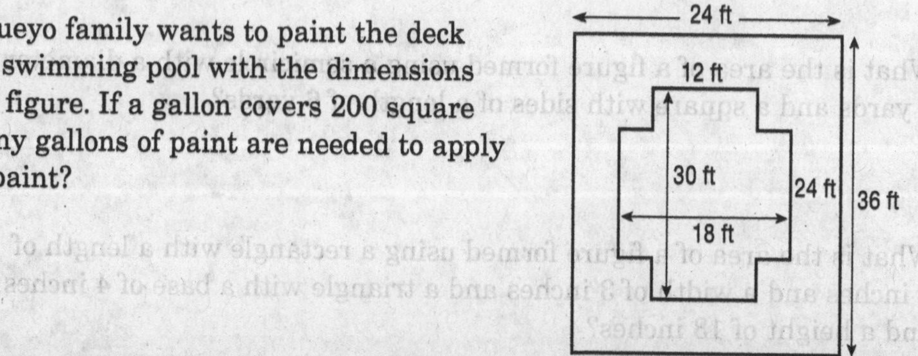

NAME ______________________ DATE ____________ PERIOD ______

7-4 Study Guide and Intervention

Three-Dimensional Figures

A **polyhedron** is a three-dimensional figure with flat surfaces that are polygons. A **prism** is a polyhedron with two parallel, congruent faces called **bases**. A **pyramid** is a polyhedron with one base that is a polygon and faces that are triangles. Prisms and pyramids are named by the shape of their bases.

Example **Identify the solid. Name the number and shapes of the faces. Then name the number of edges and vertices.**

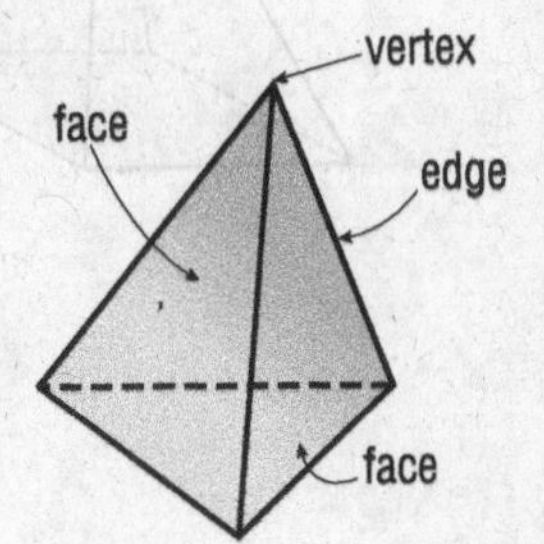

The figure has one base that is a triangle, so it is a triangular pyramid. The other faces are also triangles. It has a total of 4 faces, 6 edges, and 4 vertices.

Exercises

Identify each solid. Name the number and shapes of the faces. Then name the number of edges and vertices.

1.

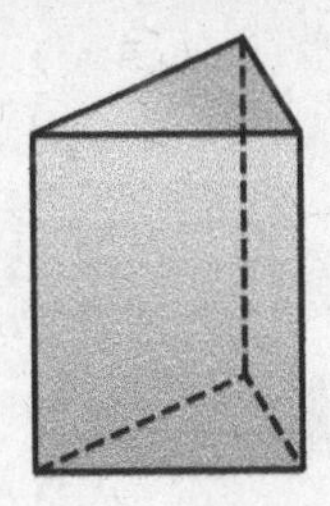

2.

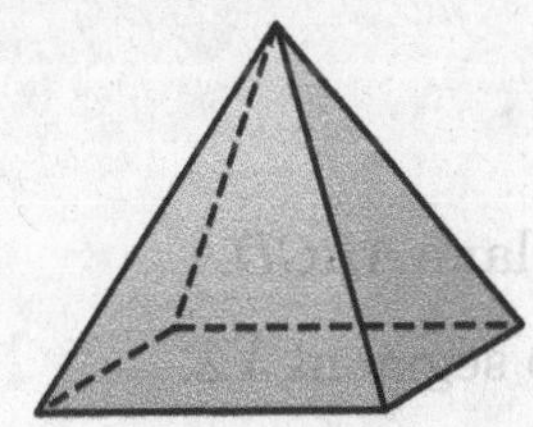

3.

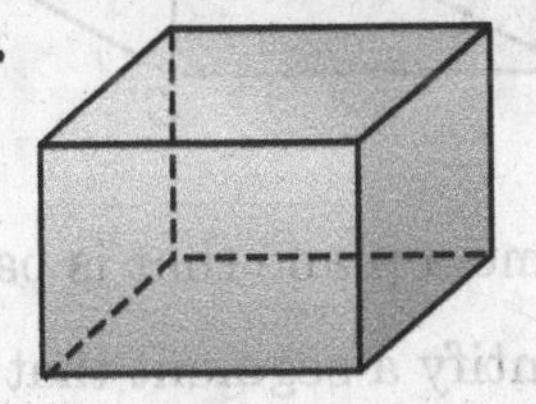

4. Draw and label the top, front, and side views of the chair shown.

NAME ______________________________ DATE ____________ PERIOD _____

7-4 Practice

Three-Dimensional Figures

Identify each solid. Name the number and shapes of the faces. Then name the number of edges and vertices.

1.

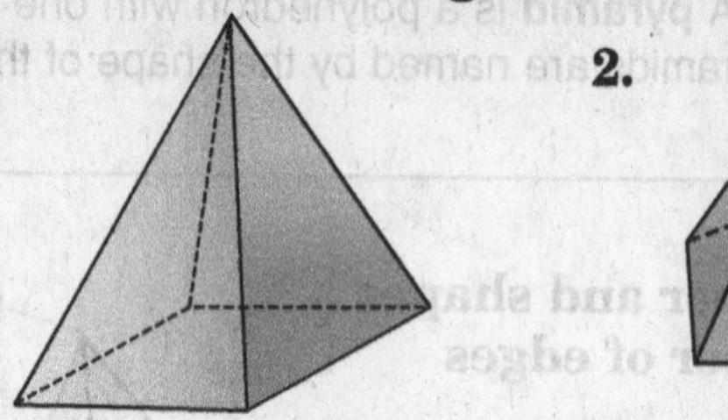

2.

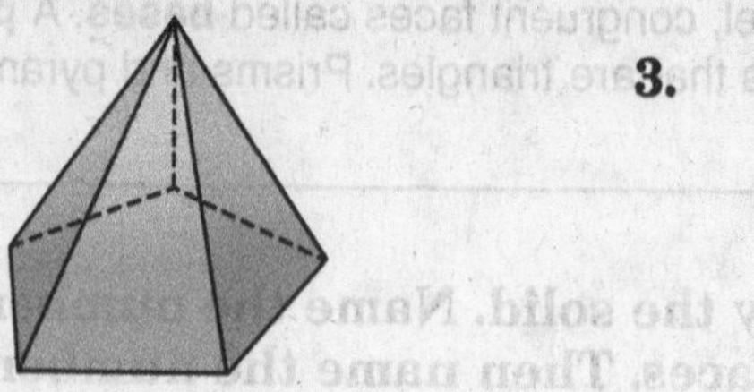

3.

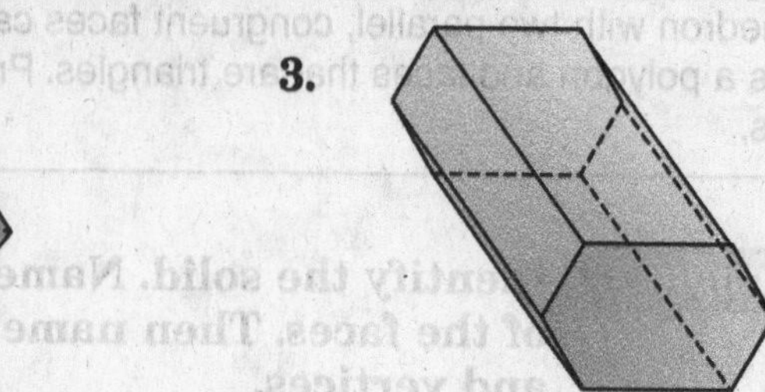

W Z A D X Y B C

4. Name a plane that is parallel to plane *ABCD*.

5. Identify a segment that is skew to segment *YZ*.

Identify each solid.

6.

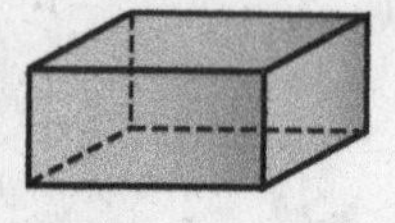

7.

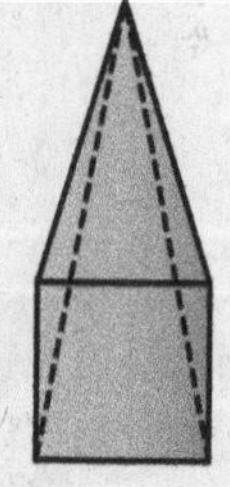

7-5 Study Guide and Intervention

Volume of Prisms and Cylinders

The volume V of a prism or a cylinder is the area of the base B times the height h, or $V = Bh$.

Example 1 **Find the volume of the rectangular prism.**

4 cm
5 cm
8 cm

$V = Bh$ — Volume of a prism
$V = (\ell \cdot w)h$ — The base is a rectangle, so $B = \ell \cdot w$.
$V = (8 \cdot 5)4$ — $\ell = 8$, $w = 5$, $h = 4$
$V = 160$ — Simplify.

The volume is 160 cubic centimeters.

The volume V of a cylinder with radius r is the area of the base B times the height h, or $V = Bh$. Since the base is a circle, the volume can also be written as $V = \pi r^2 h$, where $B = \pi r^2$.

Example 2 **Find the volume of the cylinder. Use 3.14 for π. Round to the nearest tenth if necessary.**

5 ft
25 ft

$V = \pi r^2 h$ — Volume of a cylinder
$V = 3.14 \cdot 5^2 \cdot 25$ — $\pi \approx 3.14$, $r = 5$, $h = 25$
$V \approx 1{,}962.5$ — Simplify.

The volume is about 1,962.5 cubic feet.

Exercises

Find the volume of each solid. Use 3.14 for π. Round to the nearest tenth if necessary.

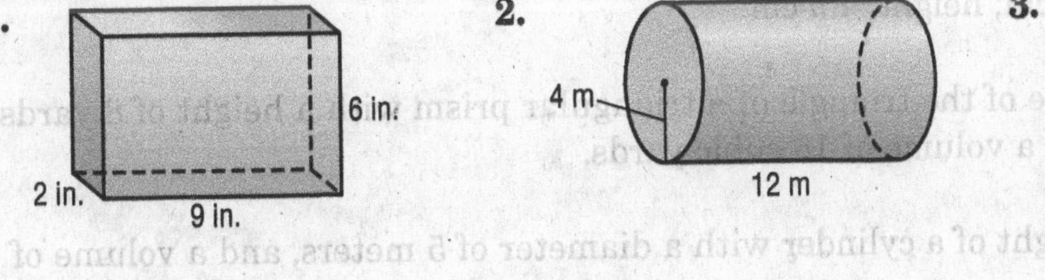

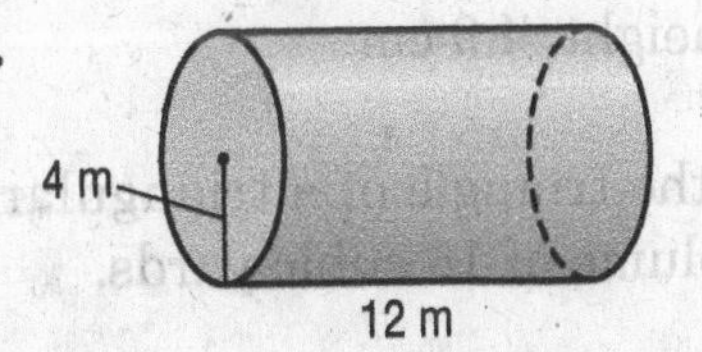

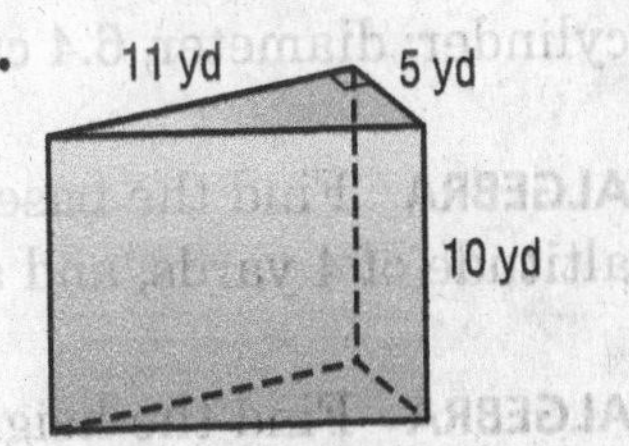

4.
3 mm
3 mm
11 mm

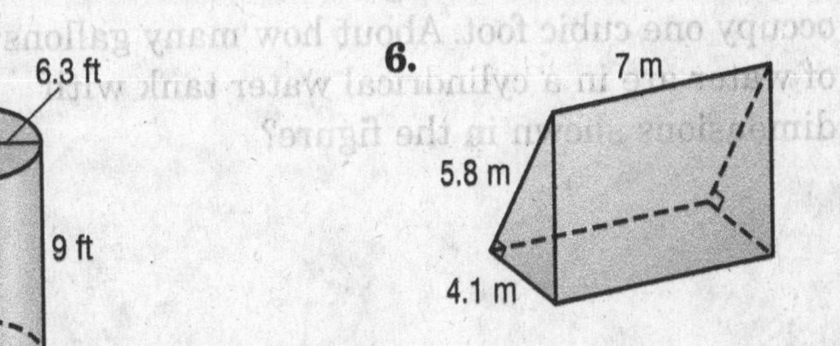

NAME ______________________________ DATE __________ PERIOD ____

7-5 Practice

Volume of Prisms and Cylinders

Find the volume of each solid. Use 3.14 for π. Round to the nearest tenth if necessary.

1.

4 m
5 m

2.

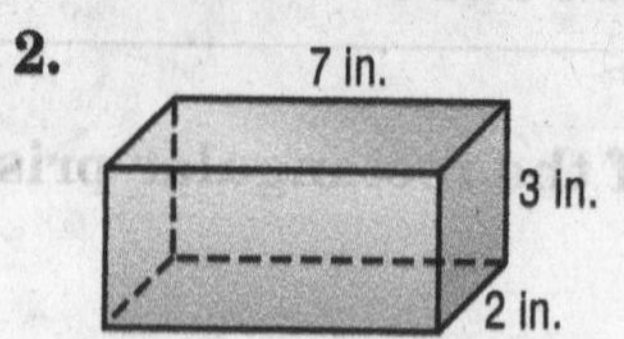

3.

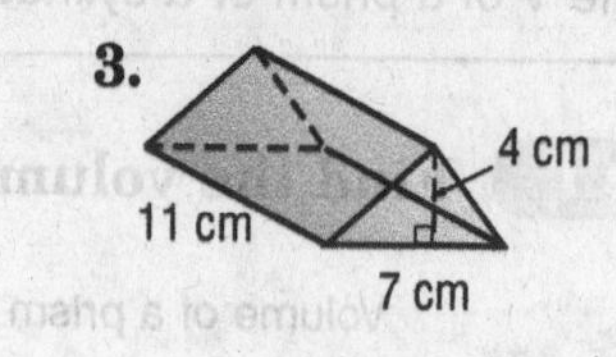

4. **5.**

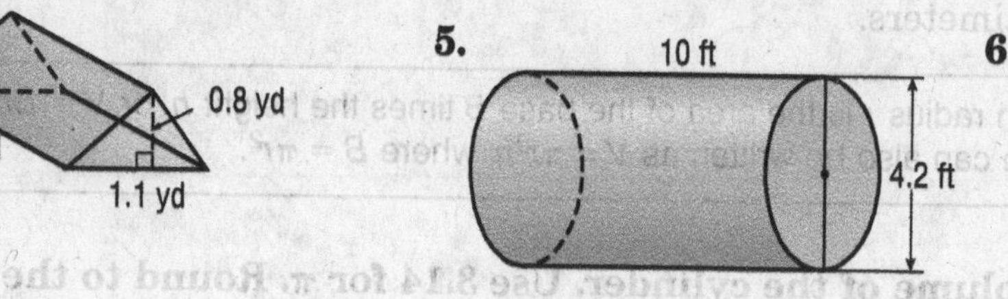

6.

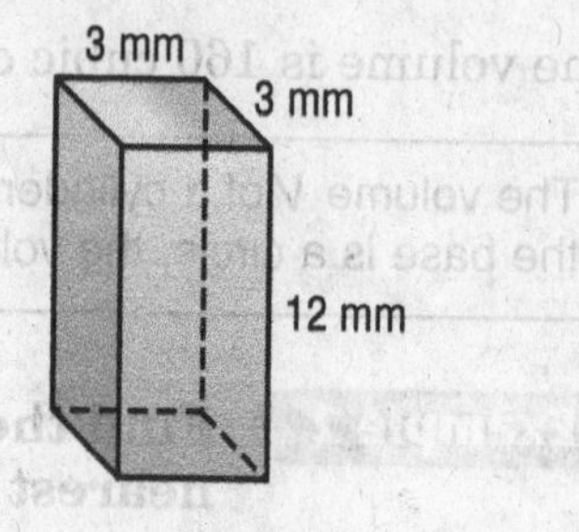

7. rectangular prism: length, 10 m; width, 5 m; height, 5 m

8. triangular prism: base of triangle, 8 in; altitude, 8 in; height of prism, 6 in

9. cylinder: radius, 7 ft; height, 4 ft

10. cylinder: diameter, 6.4 cm; height, 4.9 cm

11. ALGEBRA Find the base of the triangle of a triangular prism with a height of 8 yards, altitude of 4 yards, and a volume of 16 cubic yards.

12. ALGEBRA Find the height of a cylinder with a diameter of 5 meters, and a volume of 49.1 cubic meters.

13. WATER TANK About 7.5 gallons of water occupy one cubic foot. About how many gallons of water are in a cylindrical water tank with dimensions shown in the figure?

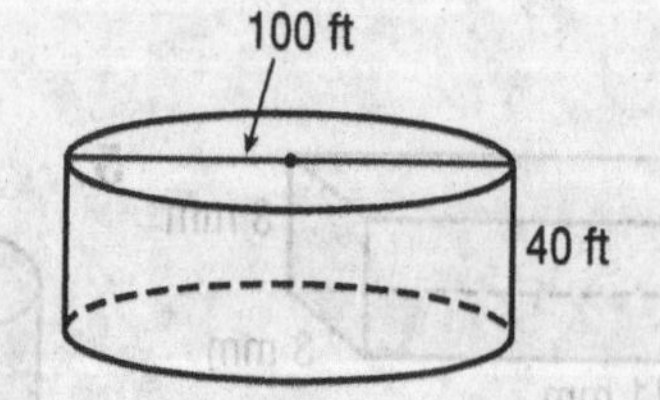

NAME ______________________ DATE ____________ PERIOD ______

7-6 Study Guide and Intervention

Volume of Pyramids and Cones

Volume Formulas	
Pyramid $V = \frac{1}{3}Bh$ V = volume, h = height, B = area of the base or ℓw	**Cone** $V = \frac{1}{3}Bh$ V = volume, h = height, B = area of the base or πr^2

Example 1 **Find the volume of the pyramid.**

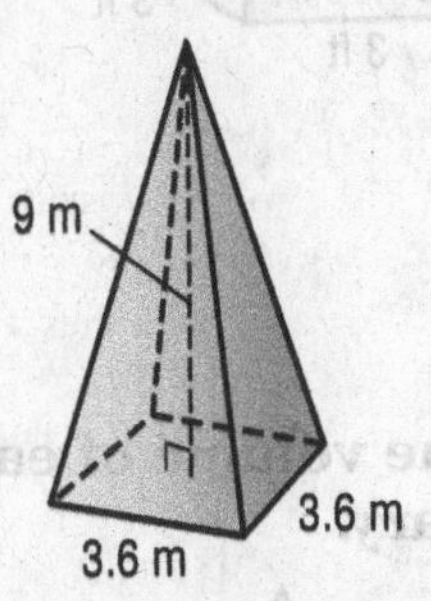

$V = \frac{1}{3}Bh$	Volume of a pyramid
$V = \frac{1}{3}s^2h$	The base is a square, so $B = s^2$.
$V = \frac{1}{3} \cdot (3.6)^2 \cdot 9$	$s = 3.6$, $h = 9$
$V = 38.88$	Simplify.

The volume is 38.88 cubic meters.

Example 2 **Find the volume of the cone. Use 3.14 for π.**

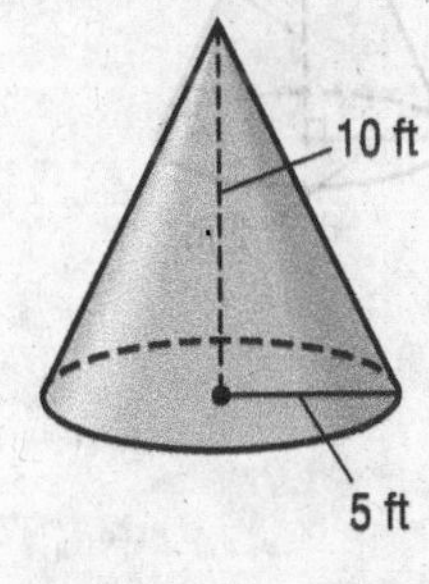

$V = \frac{1}{3}\pi r^2 h$	Volume of a cone
$V = \frac{1}{3} \cdot 3.14 \cdot 5^2 \cdot 10$	$\pi \approx 3.14$, $r = 5$, $h = 10$
$V \approx 261.7$	Simplify.

The volume is about 261.7 cubic feet.

Exercises

Find the volume of each solid. Use 3.14 for π. Round to the nearest tenth if necessary.

1.

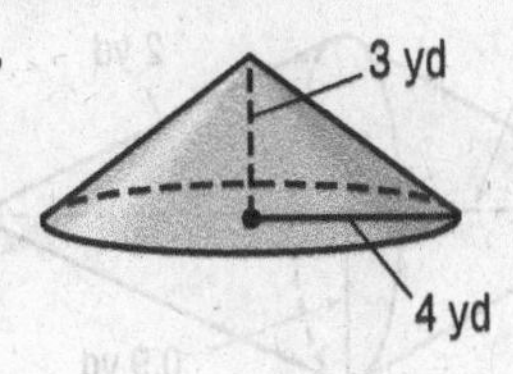

2.

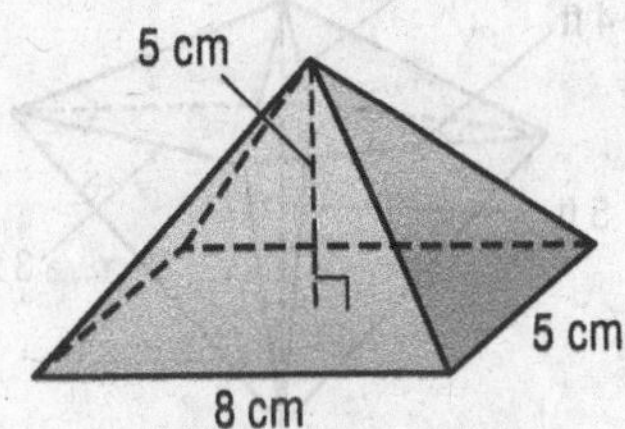

3.

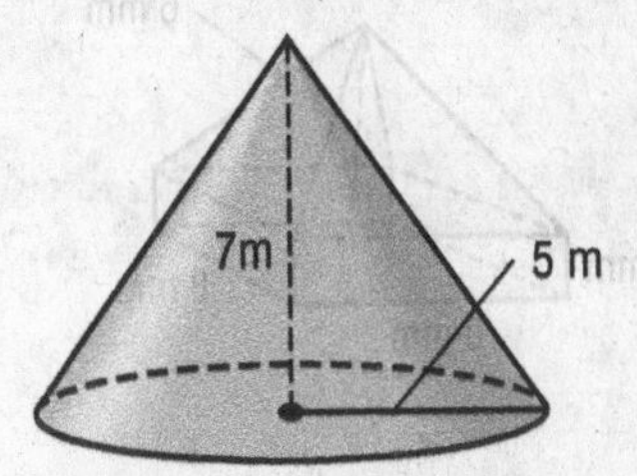

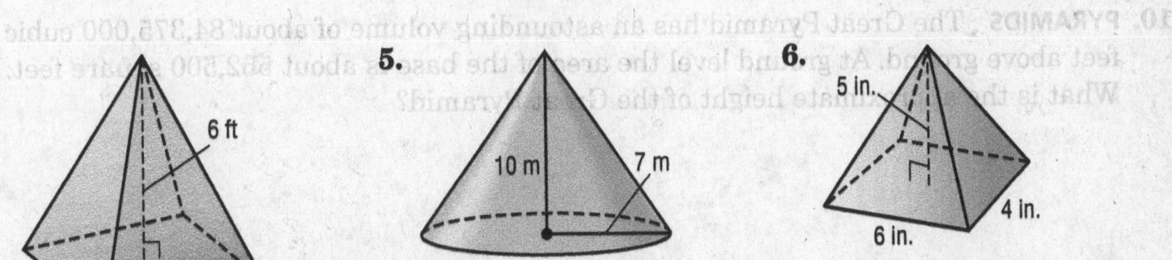

4.

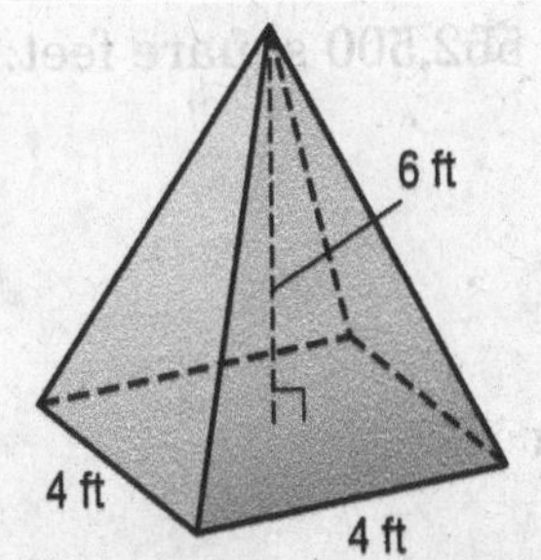

5. **6.**

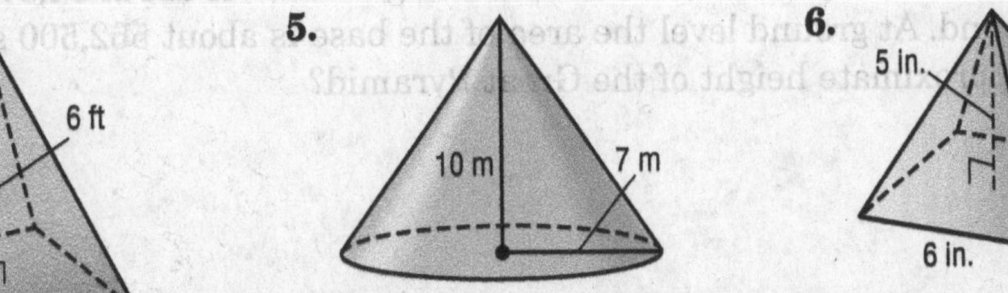

Lesson 7-6

NAME ______________________________ DATE ______________ PERIOD ______

7-6 Practice

Volume of Pyramids and Cones

Find the volume of each pyramid. Use 3.14 for π. Round to the nearest tenth if necessary.

1.

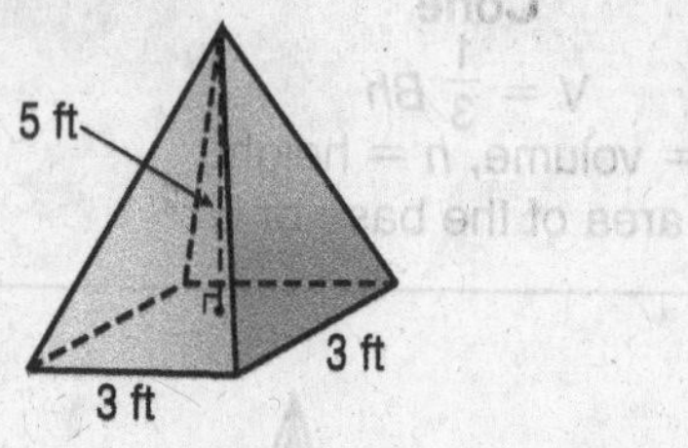

2.

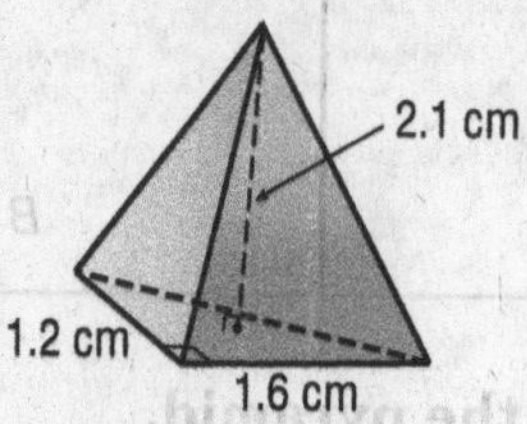

3.

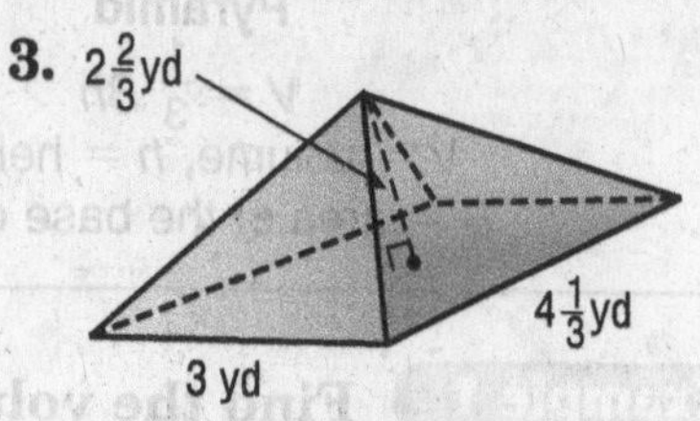

Find the volume of each cone. Use 3.14 for π. Round to the nearest tenth if necessary.

4.

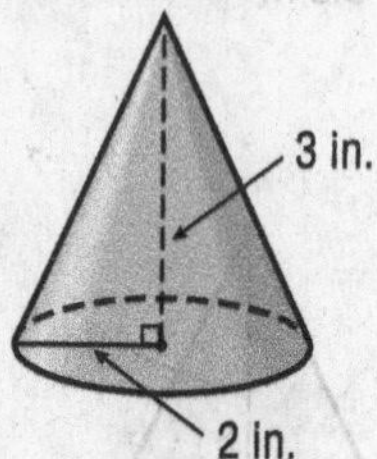

5.

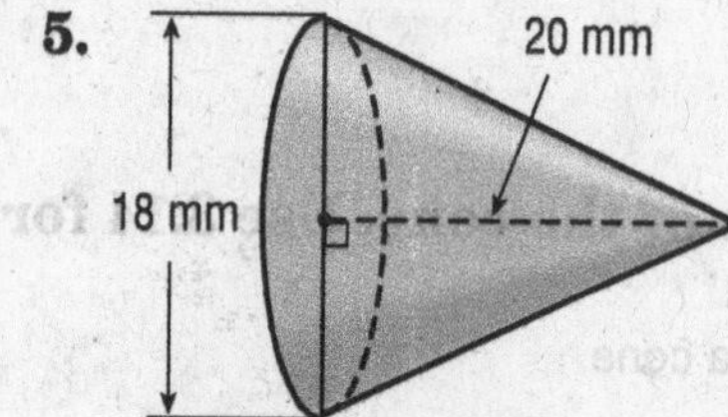

6.

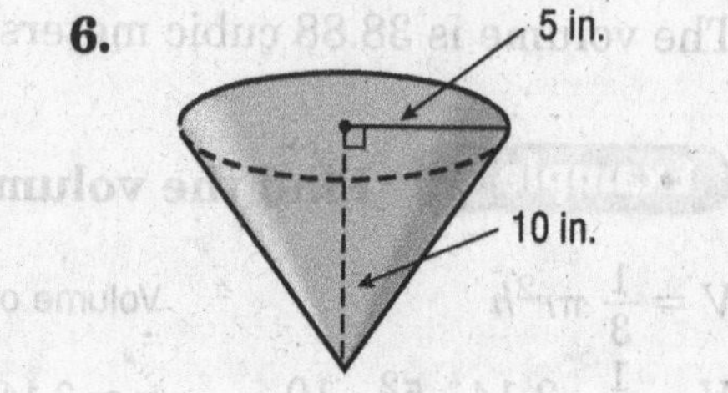

Find the volume of each solid. Use 3.14 for π. Round to the nearest tenth if necessary.

7.

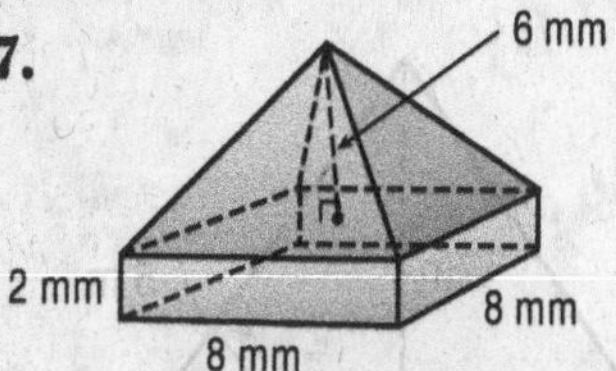

8.

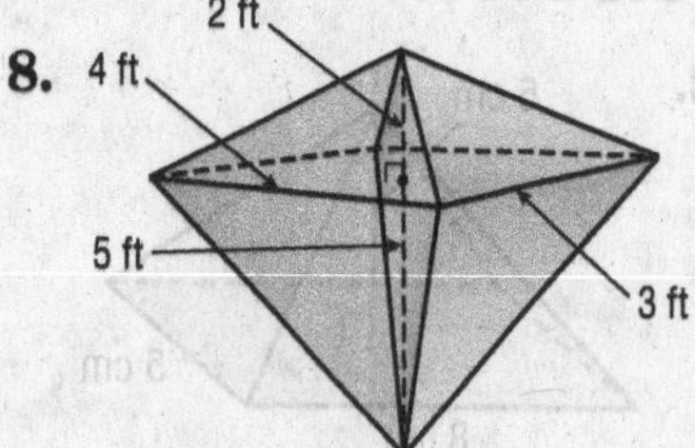

9.

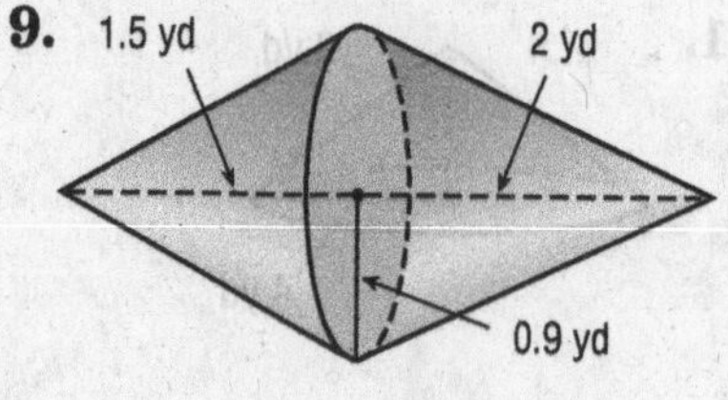

10. PYRAMIDS The Great Pyramid has an astounding volume of about 84,375,000 cubic feet above ground. At ground level the area of the base is about 562,500 square feet. What is the approximate height of the Great Pyramid?

NAME ______________________________ DATE ______________ PERIOD _____

7-7 Study Guide and Intervention

Surface Area of Prisms and Cylinders

The lateral area ℓ of a prism is the perimeter P of the base times the height h of the prism, or $\ell = Ph$. The total surface area S of a prism is the lateral surface area ℓ plus the area of the two bases $2B$, or $S = \ell + 2B$ or $S = Ph + 2B$.

Example 1 **Find the lateral and total surface areas of the rectangular prism.**

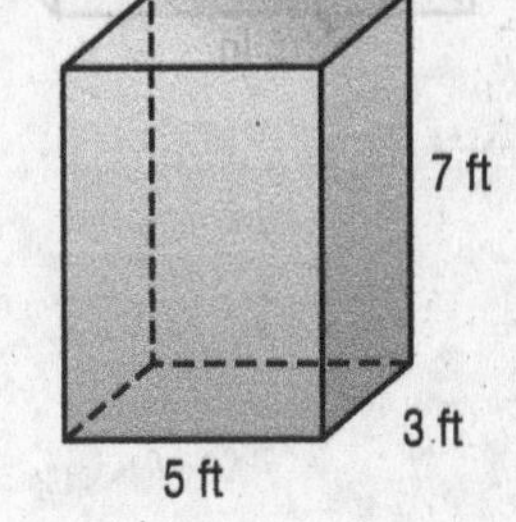

Perimeter of Base	**Area of Base**
$P = 2\ell + 2w$	$B = \ell w$
$P = 2(5) + 2(3)$	$B = 5(3)$
$P = 16$	$B = 15$

Use this information to find the lateral and total surface areas.

Lateral Surface Area	**Total Surface Area**
$L = Ph$	$S = L + 2B$
$L = 16(7)$ or 112	$S = 112 + 2(15)$ or 142

The lateral surface area is 112 square feet and the total surface area of the prism is 142 square feet.

The lateral area L of a cylinder with height h and radius r is the circumference of the base times the height, or $\ell = 2\pi rh$. The surface area S of a cylinder with height h and radius r is the lateral area plus the area of the two bases, or $S = \ell + 2\pi r^2$ or $S = 2\pi rh + 2\pi r^2$.

Example 2 **Find the surface area of the cylinder. Round to the nearest tenth.**

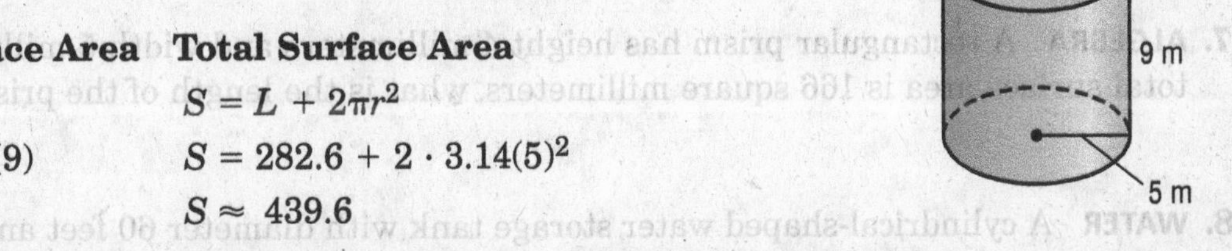

Lateral Surface Area	**Total Surface Area**
$L = 2\pi rh$	$S = L + 2\pi r^2$
$L = 2 \cdot 3.14(5)(9)$	$S = 282.6 + 2 \cdot 3.14(5)^2$
$L \approx 282.6$	$S \approx 439.6$

The lateral area is about 282.6 square meters, and the surface area of the cylinder is about 439.6 square meters.

Exercises **Find the lateral and total surface areas of each solid. Round to the nearest tenth if necessary.**

1.

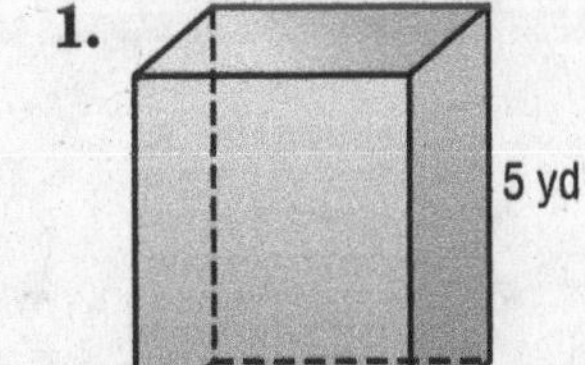

2.

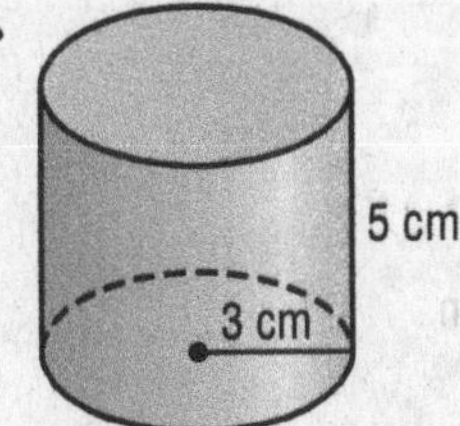

3.

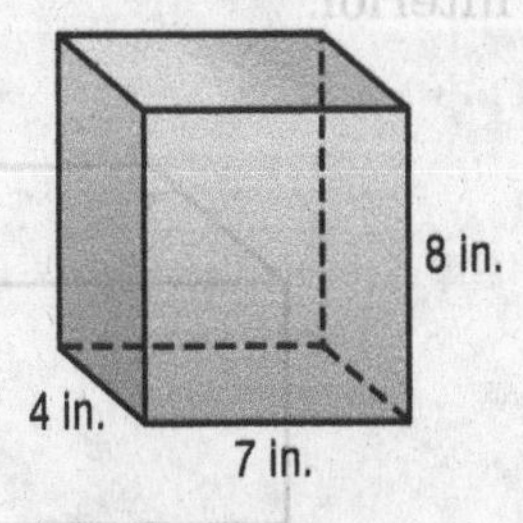

7-7 Practice

Surface Area of Prisms and Cylinders

Find the lateral and total surface areas of each solid. Use 3.14 for π. Round to the nearest tenth if necessary.

1.

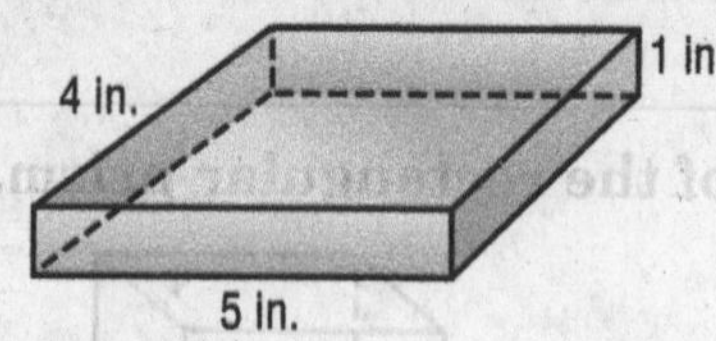

2.

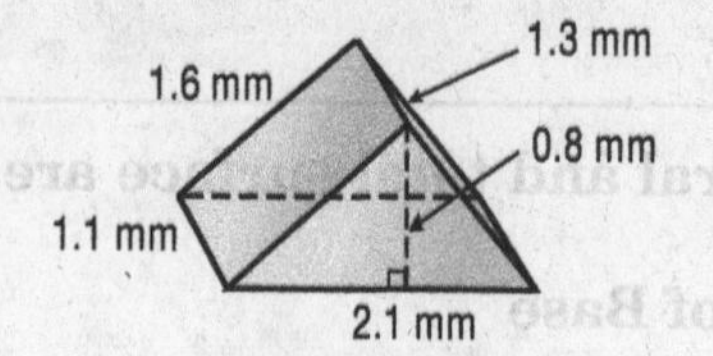

3.

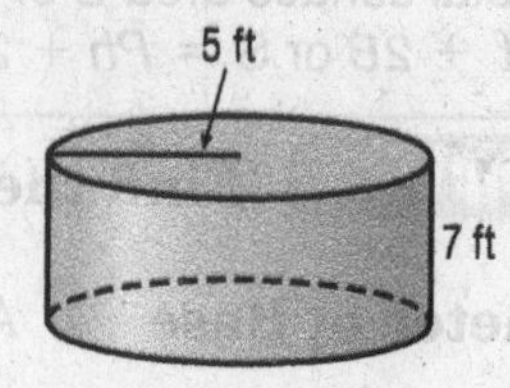

4.

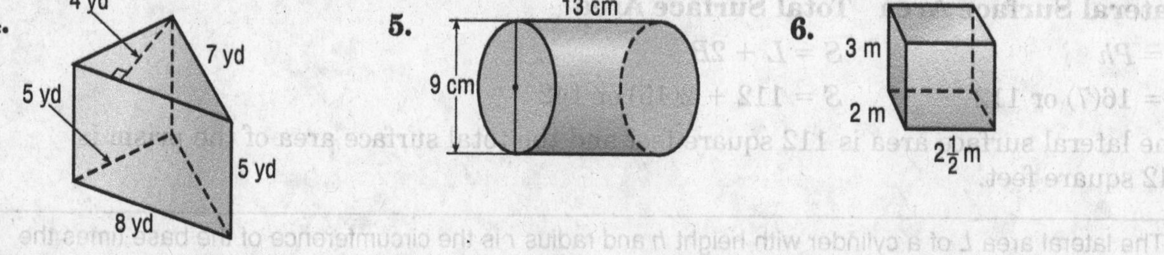

5.

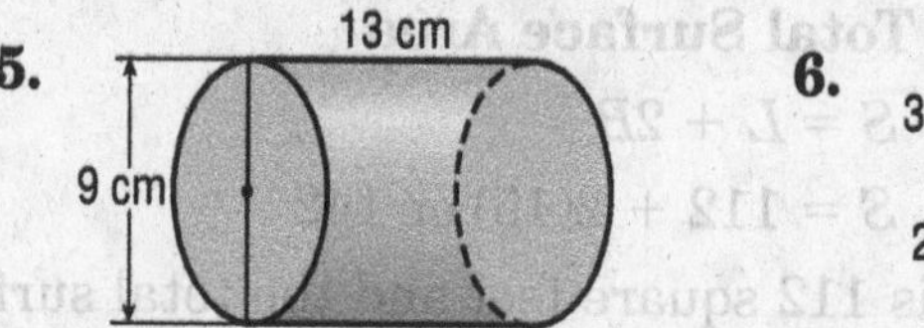

6.

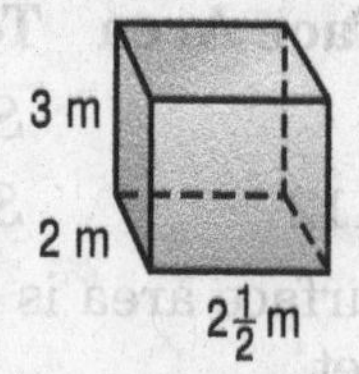

7. ALGEBRA A rectangular prism has height 4 millimeters and width 5 millimeters. If the total surface area is 166 square millimeters, what is the length of the prism?

8. WATER A cylindrical-shaped water storage tank with diameter 60 feet and height 20 feet needs to be painted on the outside. If the tank is on the ground, find the surface area that needs painting.

9. CONCRETE Find the total surface area of the hollow concrete casing shown, including the interior.

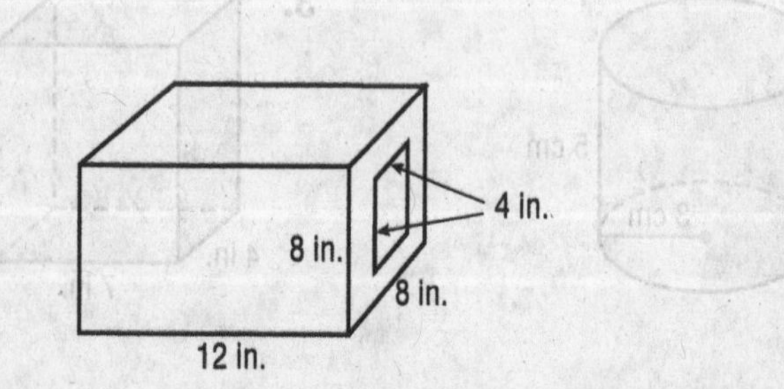

NAME ______________________________ DATE ____________ PERIOD _____

7-8 Study Guide and Intervention

Surface Area of Pyramids

The lateral surface area L of a regular pyramid is half the perimeter P of the base times the slant height ℓ, or $L = \frac{1}{2}P\ell$. The total surface areas of a regular pyramid is the lateral area L plus the area of the base B, or $S = L + B$ or $S = \frac{1}{2}P\ell + B$.

Example **Find the lateral and total surface areas of the square pyramid.**

5 ft
4 ft
4 ft

Lateral Surface Area

$L = \frac{1}{2}P\ell$

$L = \frac{1}{2}(16)(5)$ $P = 16, \ell = 5$

$L = 40$

Total Surface Area

$S = L + B$

$S = 40 + 4^2$

$S = 56$

The lateral surface area is 40 square feet, and the total surface area of the pyramid is 56 square feet.

Exercises

Find the surface area of each solid. Round to the nearest tenth if necessary.

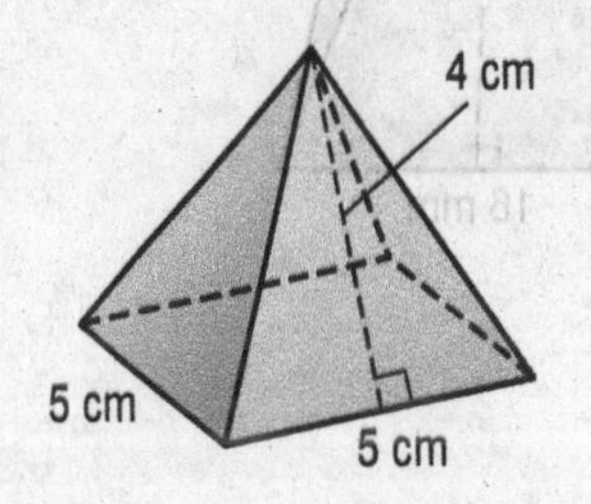

NAME ______________________ DATE __________ PERIOD _____

7-8 Practice

Surface Area of Pyramids

Find the lateral and total surface areas of each regular pyramid. Round to the nearest tenth if necessary.

1.

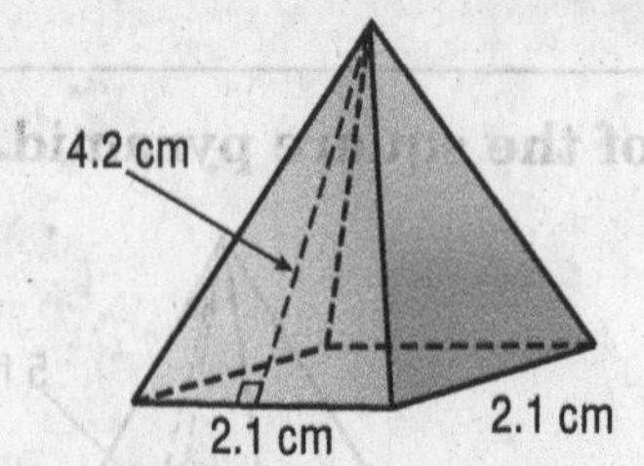

2.

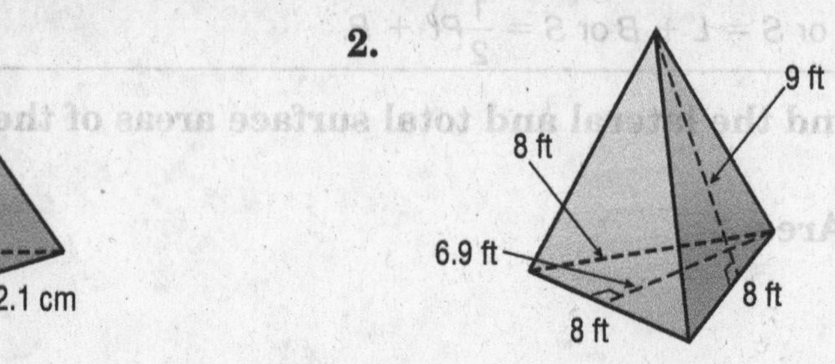

3.

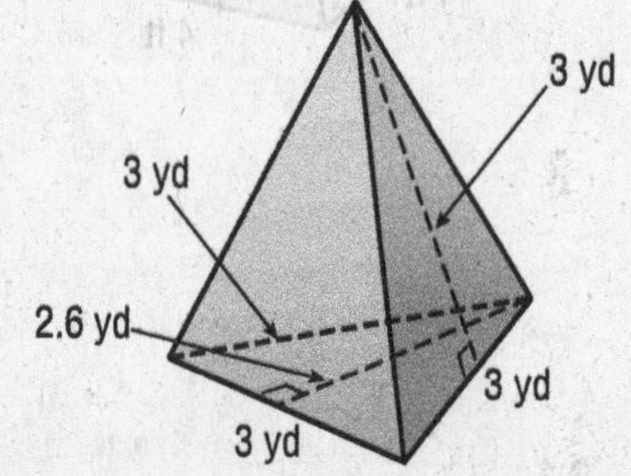

4.

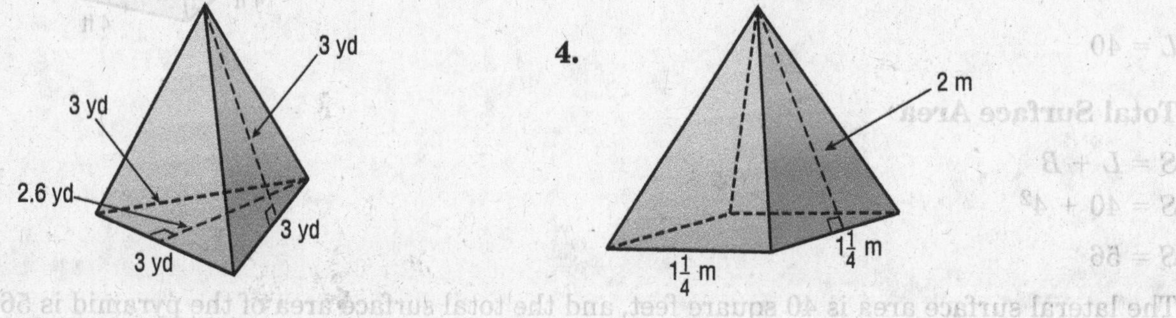

5.

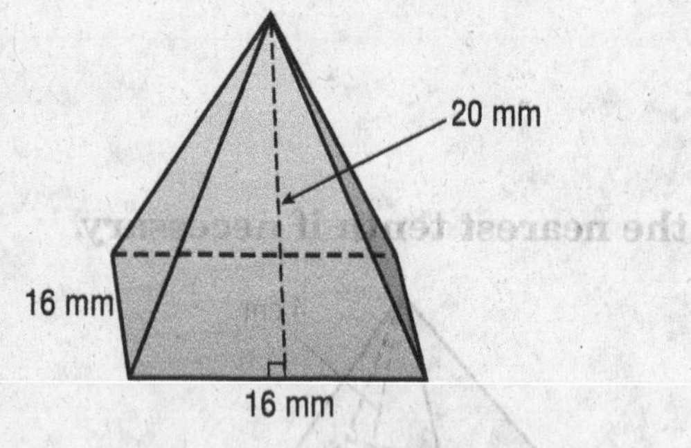

6.

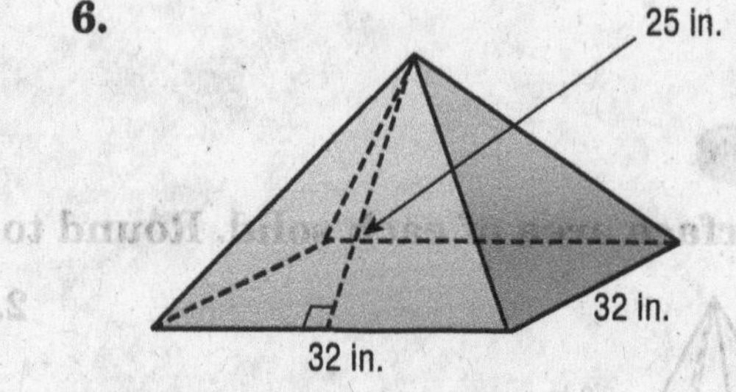

7. ALGEBRA A square pyramid has a lateral surface area of 20 square yards. If the slant height is 2 yards, what is the total surface area of the pyramid?

8. PYRAMIDS When the Great Pyramid was built, the slant height was about 610 feet and the length of the base was about 750 feet. Find the approximate lateral surface area of the Great Pyramid when it was built.

7-9 Study Guide and Intervention

Similar Solids

Similar solids have the same shape, their corresponding linear measures are proportional, and their corresponding faces are similar polygons.

Example 1 **The cones at the right are similar. Find the height of cone A.**

$\frac{8}{x} = \frac{4}{3}$	Write a ratio.
$4x = 24$	Find the cross products.
$x = 6$	Simplify.

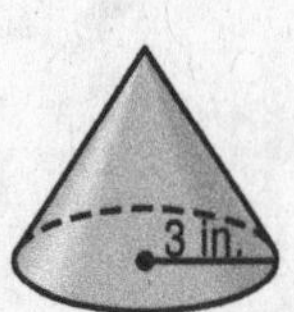

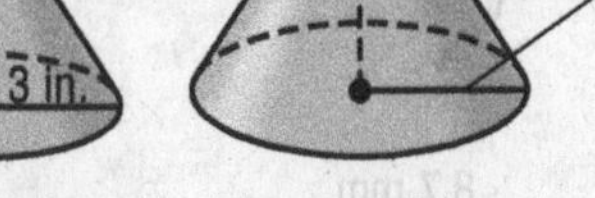

The height of the smaller cone is 6 inches.

Example 2 **The pyramids at the right are similar. Find the total surface area of pyramid B.**

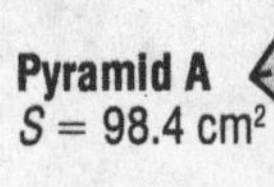

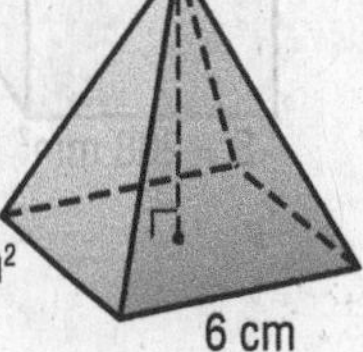

The scale factor $\frac{a}{b}$ is $\frac{6}{4}$ or $\frac{3}{2}$.

$\frac{\text{surface area of pyramid A}}{\text{surface area of pyramid B}} = \left(\frac{a}{b}\right)^2$	Write a proportion.
$\frac{98.4}{S} = \left(\frac{3}{2}\right)^2$	Substitute the known values. Let S represent the surface area.
$\frac{98.4}{S} = \frac{9}{4}$	$\left(\frac{3}{2}\right)^2 = \frac{3}{2} \cdot \frac{3}{2}$ or $\frac{9}{4}$
$98.4 \cdot 4 = 9S$	Find the cross products.
$\frac{393.6}{9} = \frac{9S}{9}$	Divide each side by 9.
$43.7 \approx S$	Simplify.

The surface area of pyramid B is approximately 43.7 square centimeters.

Exercises

For Exercises 1 and 2, the solids in each pair are similar. Find the surface area of solid B.

1.

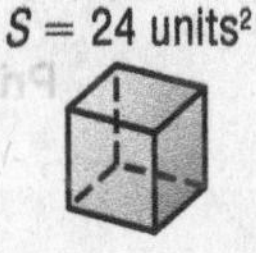

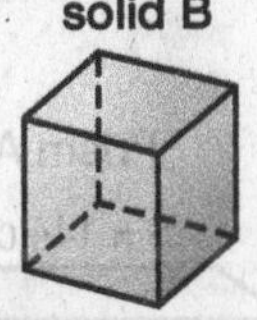

2.

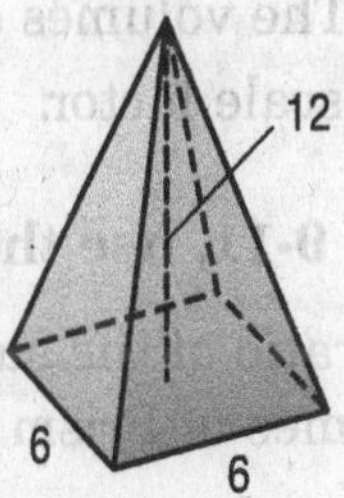

For Exercises 3 and 4, find the value of *x*.

3.

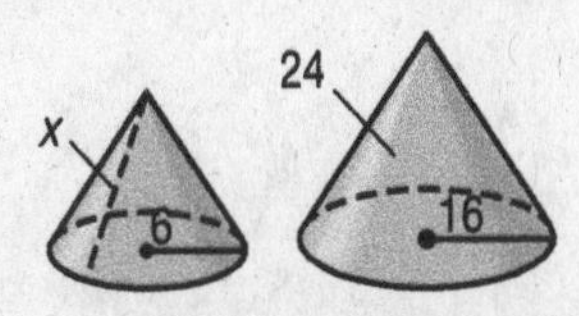

4.

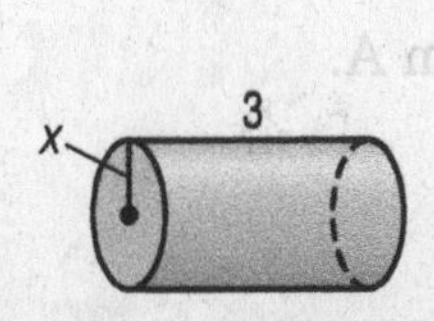

7-9 Practice

Similar Solids

Find the missing measure for each pair of similar solids. Round to the nearest tenth if necessary.

1.

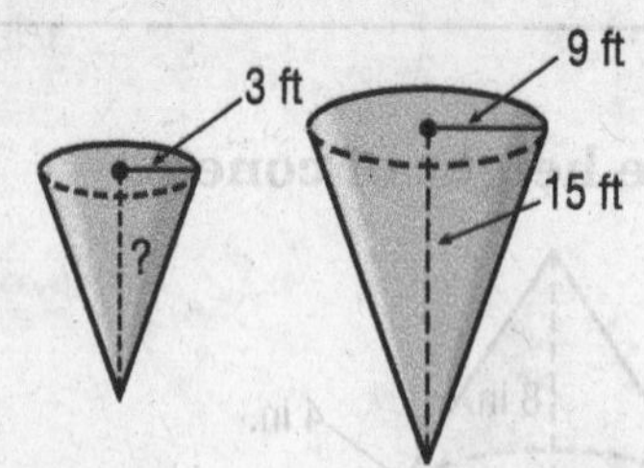

2.

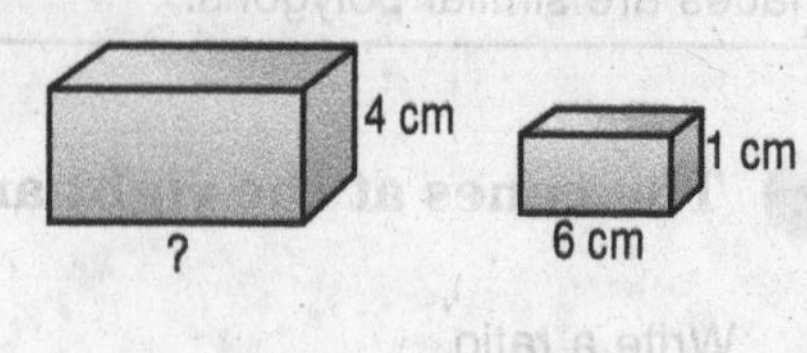

3.

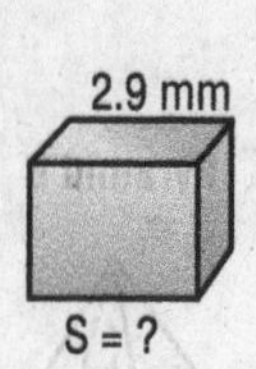

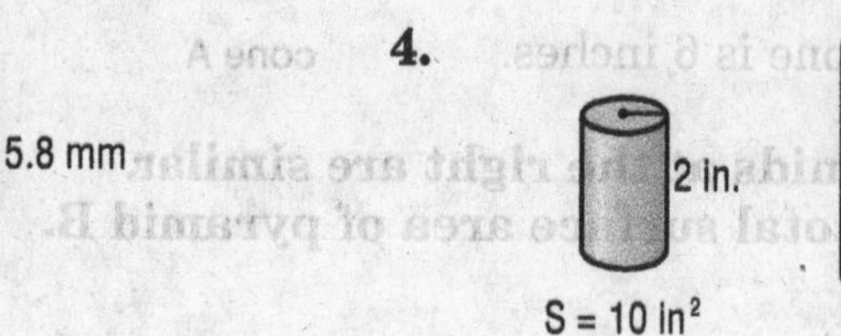

4.

1 in.
2 in.
3 in.
S = 10 in²
S = ?

5.

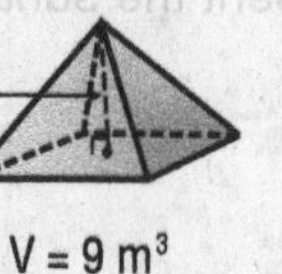

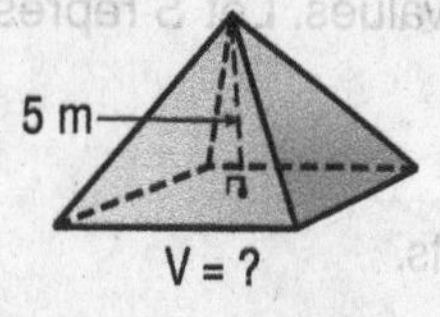

6.

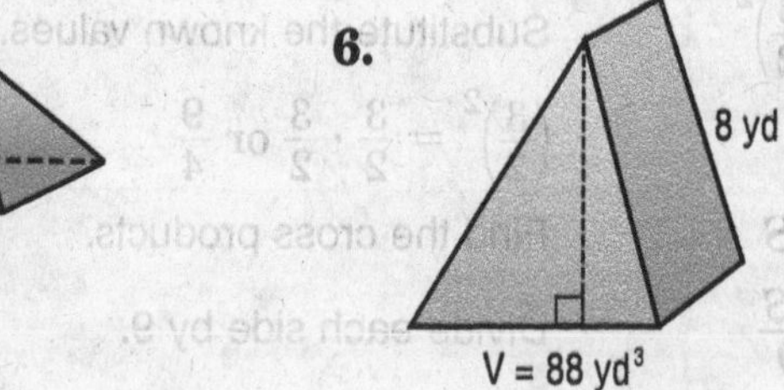

4 yd
V = ?

7. MODEL TRAINS The caboose of an N scale model train is $4\frac{1}{2}$ inches long. In the N scale, 1 inch represents $13\frac{1}{2}$ feet. What is the length of the original caboose?

8. ALGEBRA The volumes of two similar cylinders are 7 cubic meters and 56 cubic meters. Find their scale factor.

For Exercises 9-11, use the similar prisms shown.

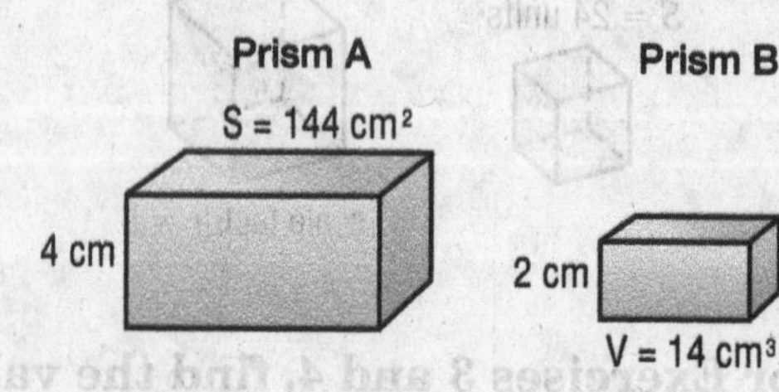

9. Write the ratio of the surface areas and the ratio of the volumes of Prism B to Prism A.

10. Find the surface area of prism B.

11. Find the volume of prism A.

NAME ________________________________ DATE ____________ PERIOD ______

8-1 Study Guide and Intervention

Simplifying Algebraic Expressions

The **Distributive Property** can be used to simplify algebraic expressions.

Examples **Use the Distributive Property to rewrite each expression.**

1 $3(a + 5)$

$3(a + 5) = 3(a) + 3(5)$ Distributive Property

$= 3a + 15$ Simplify.

2 $-2(d - 3)$

$-2(d - 3) = -2[d + (-3)]$ Rewrite $d - 3$ as $d + (-3)$.

$= -2(d) + (-2)(-3)$ Distributive Property

$= -2(d) + 6$ Simplify.

When a plus sign separates an algebraic expression into parts, each part is called a **term**. In terms that contain a variable, the numerical part of the term is called the **coefficient** of the variable. A term without a variable is called a **constant. Like terms** contain the same variables, such as $3x$ and $2x$.

Example 3 **Identify the terms, like terms, coefficients, and constants in the expression $7x - 5 + x - 3x$.**

$7x - 5 + x - 3x = 7x + (-5) + x + (-3x)$ Definition of subtraction

$= 7x + (-5) + 1x + (-3x)$ Identity Property; $x = 1x$

The terms are $7x$, -5, x, and $-3x$. The like terms are $7x$, x, and $-3x$. The coefficients are 7, 1, and -3. The constant is -5.

An algebraic expression is in **simplest form** if it has no like terms and no parentheses.

Example 4 **Simplify the expression $-2m + 5 + 6m - 3$.**

$-2m$ and $6m$ are like terms. 5 and -3 are also like terms.

$-2m + 5 + 6m - 3 = -2m + 5 + 6m + (-3)$ Definition of subtraction

$= -2m + 6m + 5 + (-3)$ Commutative Property

$= (-2 + 6)m + 5 + (-3)$ Distributive Property

$= 4m + 2$ Simplify.

Exercises

Use the Distributive Property to rewrite each expression.

1. $2(c + 6)$ **2.** $-4(w + 6)$ **3.** $(b - 4)(-3)$

4. Identify the terms, like terms, coefficients, and constants in the expression $4m - 2 + 3m + 5$.

Simplify each expression.

5. $3d + 6d$ **6.** $2 + 5s - 4$ **7.** $2z + 3 + 9z - 8$

Lesson 8-1

NAME ______________________________ DATE ____________ PERIOD _____

8-1 Practice

Simplifying Algebraic Expressions

Use the Distributive Property to rewrite each expression.

1. $6(z + 4)$ **2.** $-7(c + 2)$ **3.** $(d + 5)9$ **4.** $(h + 8)(-3)$

5. $5(y - 2)$ **6.** $3(6 - n)$ **7.** $-4(s - 4)$ **8.** $-9(2 - p)$

9. $2(3x + 1)$ **10.** $-5(4n - 5)$ **11.** $8(u - 2v)$ **12.** $3a(7b + 6c)$

Identify the terms, like terms, coefficients, and constants in each expression.

13. $4b + 7b + 5$ **14.** $8 + 6t - 3t + t$ **15.** $-5x + 4 - x - 1$

Simplify each expression.

16. $h + 6h$ **17.** $10k - k$ **18.** $3b + 8 + 2b$

19. $4 + 5v + v$ **20.** $-2f + 3 - 2f - 8$ **21.** $-7s - 5 - 7s + 9$

22. $-\frac{3}{4}x - \frac{1}{3} + \frac{7}{8}x - \frac{1}{2}$ **23.** $5c - 3d - 12c + d$ **24.** $-y + 9z - 16y - 25z$

Write two equivalent expressions for the area of each figure.

25.

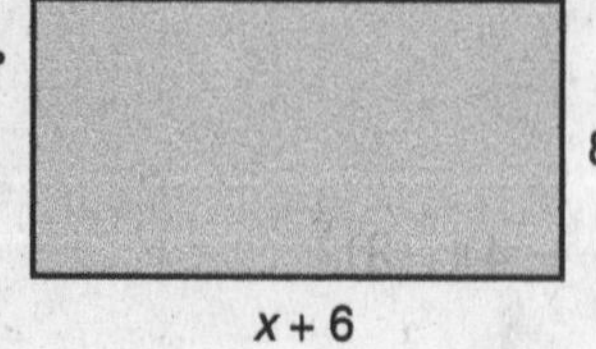

26.

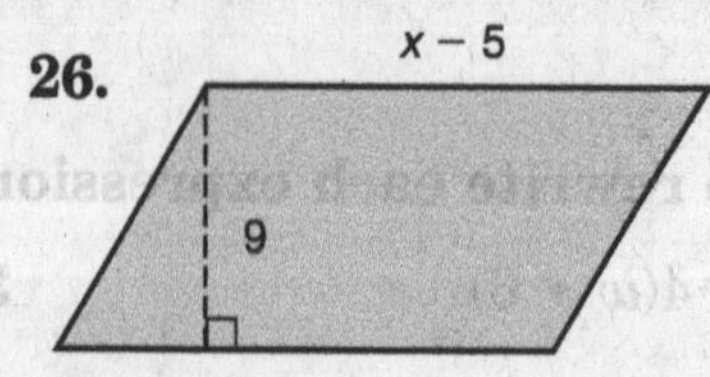

27.

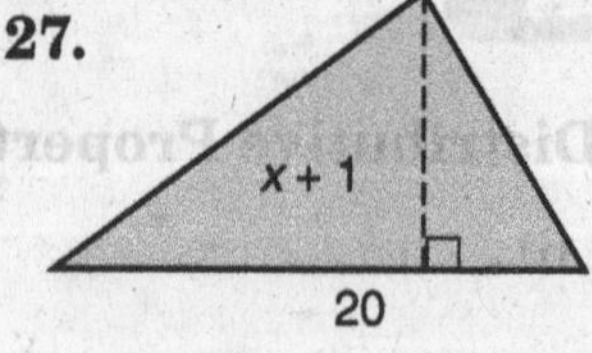

28. PAINTING Mr. Torres paid $43 for supplies to paint his office. He paid one person $8 per hour to prepare the office to be painted and another person $10 per hour to paint the office. If both people worked h hours, write two expressions that you could use to represent the total cost of painting the office.

NAME ____________________ DATE __________ PERIOD ______

8-2 Study Guide and Intervention

Solving Two-Step Equations

A **two-step equation** contains two operations. To solve a two-step equation, undo each operation in reverse order.

Example 1 **Solve $-2a + 6 = 14$. Check your solution.**

Method 1 Vertical Method

$-2a + 6 = 14$ Write the equation.

$\underline{\quad -6 = -6}$ Subtract 6 from each side.

$-2a = 8$ Simplify.

$\frac{-2a}{-2} = \frac{8}{-2}$ Divide each side by -2.

$a = -4$ Simplify.

Method 2 Horizontal Method

$-2a + 6 = 14$

$-2a + 6 - 6 = 14 - 6$

$-2a = 8$

$\frac{-2a}{-2} = \frac{8}{-2}$

$a = -4$

Check $-2a + 6 = 14$ Write the equation.

$-2(-4) + 6 \stackrel{?}{=} 14$ Replace a with -4 to see if the sentence is true.

$14 = 14$ ✓ The sentence is true.

The solution is -4.

Sometimes it is necessary to combine like terms before solving an equation.

Example 2 **Solve $5 = 8x - 2x - 7$. Check your solution.**

$5 = 8x - 2x - 7$ Write the equation.

$5 = 6x - 7$ Combine like terms.

$5 + 7 = 6x - 7 + 7$ Add 7 to each side.

$12 = 6x$ Simplify.

$\frac{12}{6} = \frac{6x}{6}$ Divide each side by 6.

$2 = x$ Simplify.

The solution is 2. Check this solution.

Exercises

Solve each equation. Check your solution.

1. $2d + 7 = 9$ **2.** $11 = 3z + 5$ **3.** $2s - 4 = 6$

4. $-12 = 5r + 8$ **5.** $-6p - 3 = 9$ **6.** $-14 = 3x + x - 2$

7. $5c + 2 - 3c = 10$ **8.** $3 + 7n + 2n = 21$ **9.** $21 = 6r + 5 - 7r$

10. $8 - 5b = -7$ **11.** $-10 = 6 - 4m$ **12.** $-3t + 4 = 19$

13. $2 + \frac{a}{6} = 5$ **14.** $-\frac{1}{3}q - 7 = -3$ **15.** $4 - \frac{v}{5} = 0$

Lesson 8-2

NAME ______________________ DATE __________ PERIOD _____

8-2 Practice

Solving Two-Step Equations

Solve each equation. Check your solution.

1. $3g + 5 = 17$

2. $9 = 4a + 13$

3. $13 = 5m - 2$

4. $-15 = 2t - 11$

5. $7k - 5 = -19$

6. $13 = 4x - 11$

7. $10 = \frac{z}{2} + 7$

8. $6 + \frac{n}{5} = -4$

9. $4 - 3y = 31$

10. $15 - 2b = -9$

11. $-\frac{1}{3}y - 6 = -11$

12. $16 - \frac{r}{7} = 21$

13. $30 = 5d - 8d$

14. $w + 3w = 20$

15. $5 - 7m + 9m = 11$

16. $-18 = 8x - 9 - 5x$

17. $25 = s + 13 - 4s$

18. $6a + 7 - a = -18$

19. $3(y + 5) = 21$

20. $7(p - 3) = 35$

21. $-48 = 6(v + 2)$

22. $\frac{k - 3}{4} = 10$

23. $\frac{z + 5}{7} = -3$

24. $\frac{9 + t}{12} = -3$

25. SHOPPING Mrs. Williams shops at a store that has an annual membership fee of $30. Today she paid her annual membership and bought several fruit baskets costing $15 each as gifts for her coworkers. Her total was $105. Solve the equation $15b + 30 = 105$ to find the number of fruit baskets Mrs. Williams purchased.

26. GAMES A card game has 50 cards. After dealing 7 cards to each player, Tupi has 15 cards left over. Solve the equation $50 - 7p = 15$ to find the number of players.

27. GEOMETRY Write an equation to represent the length of $\overline{PQ}$. Then find the value of y.

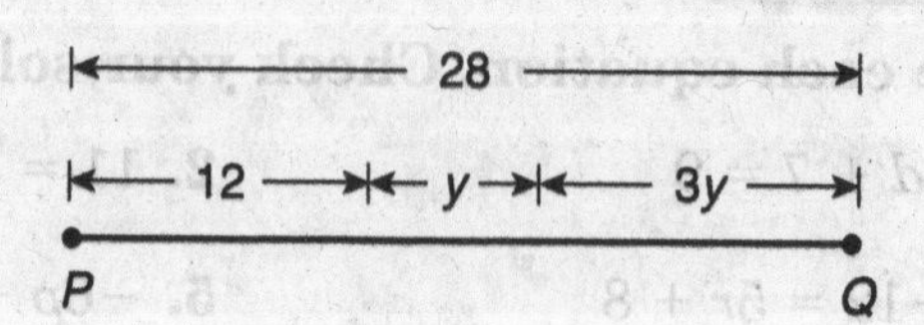

NAME ________________ DATE ________ PERIOD ____

8-3 Study Guide and Intervention

Writing Two-Step Equations

Some verbal sentences translate to two-step equations.

Example 1 **Translate each sentence into an equation.**

Sentence	Equation
Four more than three times a number is 19.	$3n + 4 = 19$
Five is seven less than twice a number.	$5 = 2n - 7$
Seven more than the quotient of a number and 3 is 10.	$7 + \frac{n}{3} = 10$

After a sentence has been translated into a two-step equation, you can solve the equation.

Example 2 **Translate the sentence into an equation. Then find the number. Thirteen more than five times a number is 28.**

Words Thirteen more than five times a number is 28.

Variable Let n = the number.

Equation

$5n + 13 = 28$	Write the equation.
$5n + 13 - 13 = 28 - 13$	Subtract 13 from each side.
$5n = 15$	Simplify.
$\frac{5n}{5} = \frac{15}{5}$	Divide each side by 5.
$n = 3$	Simplify.

Therefore, the number is 3.

Exercises

Translate each sentence into an equation. Then find each number.

1. Five more than twice a number is 7.

2. Fourteen more than three times a number is 2.

3. Seven less than twice a number is 5.

4. Two more than four times a number is -10.

5. Eight less than three times a number is -14.

6. Three more than the quotient of a number and 2 is 7.

NAME ______________________ DATE __________ PERIOD _____

8-3 Practice

Writing Two-Step Equations

Translate each sentence into an equation.

1. Three more than eight times a number is equal to 19.

2. Twelve less than seven times a number is 16.

3. Four more than twice a number is -10.

4. Nine less than five times a number is equal to -30.

5. **ART** Ishi bought a canvas and 8 tubes of paint for $24.95. If the canvas cost $6.95, how much did each tube of paint cost?

6. **ENGINEERING** The world's two highest dams are both in Tajikistan. The Rogun dam is 35 meters taller than the Nurek dam. Together they are 635 meters tall. Find the height of the Nurek dam.

U.S. PRESIDENTS For Exercises 7 and 8, use the information at the right.

President	Age at First Inauguration
J. Carter	52
R. Reagan	69
G. H. W. Bush	
W. Clinton	46
G. W. Bush	54

7. If you double President Reagan's age at the time of his first inauguration and subtract his age at the time he died, the result is 45 years. How old was President Reagan when he died?

8. If you divide the age of the first President Bush when he was inaugurated by 2 and add 14 years, you get the age of President Clinton when he was first inaugurated. How old was President G. H. W. Bush when he was inaugurated?

9. **GEOMETRY** Find the value of x in the triangle at the right.

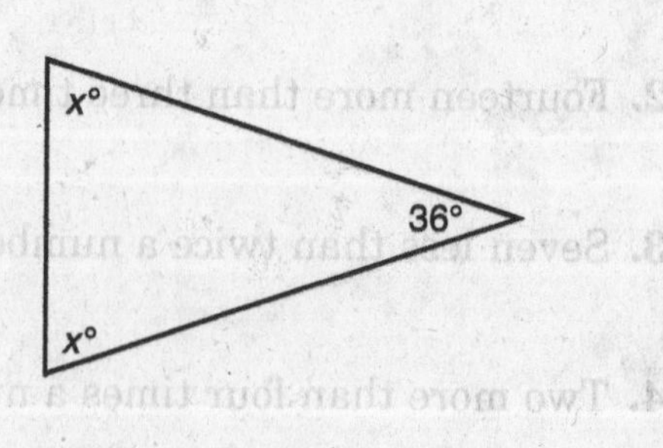

10. **ALGEBRA** Three consecutive integers can be represented by n, $n + 1$, and $n + 2$. If the sum of three consecutive integers is 57, what are the integers?

8-4 Study Guide and Intervention

Solving Equations with Variables on Each Side

Some equations, such as $3x - 9 = 6x$, have variables on each side of the equals sign. Use the Addition or Subtraction Property of Equality to write an equivalent equation with the variables on one side of the equals sign. Then solve the equation.

Example 1 **Solve $3x - 9 = 6x$. Check your solution.**

$3x - 9 = 6x$	Write the equation.
$3x - 3x - 9 = 6x - 3x$	Subtract 3*x* from each side.
$-9 = 3x$	Simplify.
$-3 = x$	Mentally divide each side by 3.

To check your solution, replace x with -3 in the original equation.

Check $3x - 9 = 6x$	Write the equation.
$3(-3) - 9 \stackrel{?}{=} 6(-3)$	Replace *x* with -3.
$-18 = -18$ ✓	The sentence is true.

The solution is -3.

Example 2 **Solve $4a - 7 = 5 - 2a$.**

$4a - 7 = 5 - 2a$	Write the equation.
$4a + 2a - 7 = 5 - 2a + 2a$	Add 2*a* to each side.
$6a - 7 = 5$	Simplify.
$6a - 7 + 7 = 5 + 7$	Add 7 to each side.
$6a = 12$	Simplify.
$a = 2$	Mentally divide each side by 6.
The solution is 2.	Check this solution.

Exercises

Solve each equation. Check your solution.

1. $6s - 10 = s$ **2.** $8r = 4r - 16$ **3.** $25 - 3u = 2u$

4. $14t - 8 = 6t$ **5.** $k + 20 = 9k - 4$ **6.** $11m + 13 = m + 23$

7. $-4b - 5 = 3b + 9$ **8.** $6y - 1 = 27 - y$ **9.** $1.6h - 72 = 4h - 30$

10. $8.5 - 3z = -8z$ **11.** $10x + 8 = 5x - 3$ **12.** $16 - 7d = -3d + 2$

NAME ______________________________ DATE ____________ PERIOD _____

8-4 Practice

Solving Equations with Variables on Each Side

Solve each equation. Check your solution.

1. $9m + 14 = 2m$
2. $13x = 32 + 5x$
3. $8d - 25 = 3d$
4. $t - 27 = 4t$
5. $7p - 5 = 6p + 8$
6. $11z - 5 = 9z + 7$
7. $12 - 5h = h + 6$
8. $4 - 7f = f - 12$
9. $-6y + 17 = 3y - 10$
10. $3x - 32 = -7x + 28$
11. $3.2a - 16 = 4a$
12. $16.8 - v = 6v$

Find each number.

13. Fourteen less than five times a number is three times the number. Define a variable, write an equation, and solve to find the number.

14. Twelve more than seven times a number equals the number less six. Define a variable, write an equation, and solve to find the number.

Write an equation to find the value of *x* so that each pair of polygons has the same perimeter. Then solve.

15.

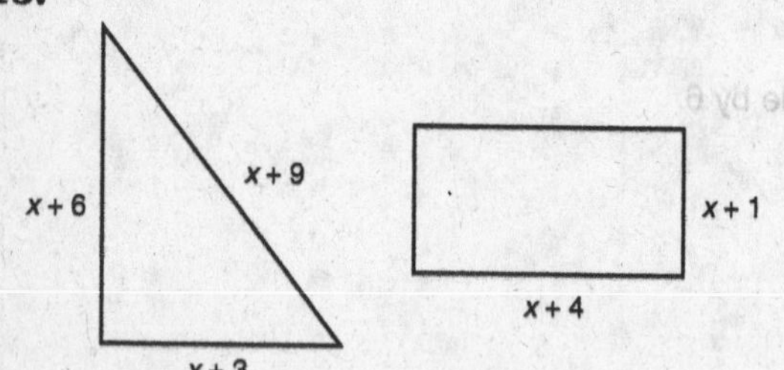

16.

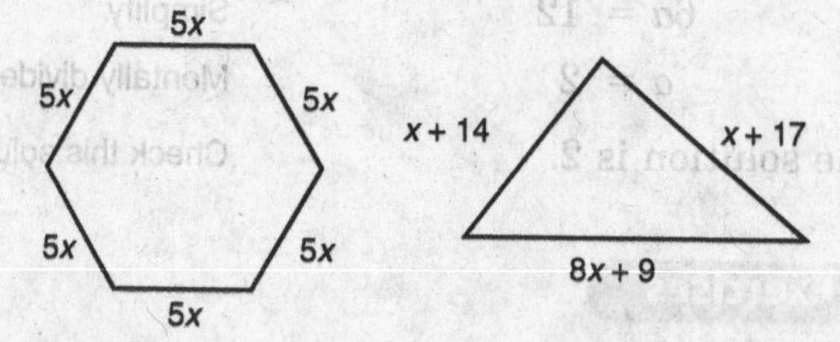

17. **GOLF** For an annual membership fee of $500, Mr. Bailey can join a country club that would allow him to play a round of golf for $35. Without the membership, the country club charges $55 for each round of golf. Write and solve an equation to determine how many rounds of golf Mr. Bailey would have to play for the cost to be the same with and without a membership.

18. **MUSIC** Marc has 45 CDs in his collection, and Andrea has 61. If Marc buys 4 new CDs each month and Andrea buys 2 new CDs each month, after how many months will Marc and Andrea have the same number of CDs?

NAME ______________________ DATE __________ PERIOD _____

8-5 Study Guide and Intervention

Problem-Solving Investigation: Guess and Check

You may need to use the guess and check strategy to solve some problems.

Understand • Determine what information is given in the problem and what you need to find.

Plan • Select a strategy including a possible estimate.

Solve • Solve the problem by carrying out your plan.

Check • Examine your answer to see if it seems reasonable.

Example

The school booster club spent $776 on ski passes for the school ski trip. Adult tickets cost $25 each and student tickets cost $18 each. They bought four times as many student tickets as adult tickets. Find the number of adult and student tickets purchased.

Understand Adult tickets cost $25 each and student tickets cost $18 each. They bought four times more student tickets than adult tickets. The total amount paid for the tickets was $776.

Plan Make a guess and check to see if it is correct. Remember, the number you guess for the student tickets must be four times more than the number you guess for adult tickets.

Solve You need to find the combination that gives a total of $776. Make a list and use a to represent the number of adult tickets and s to represent the number of student tickets.

Guess	$\$25a + \$18s = \$776$	Check
If $a = 10$, then $s = 4(10) = 40$	$\$25(10) + \$18(40) = \$970$	too high
If $a = 5$, then $s = 4(5) = 20$	$\$25(5) + \$18(20) = \$485$	too low
If $a = 7$, then $s = 4(7) = 28$	$\$25(7) + \$18(28) = \$679$	still too low
If $a = 8$, then $s = 4(8) = 32$	$\$25(8) + \$18(32) = \$776$	correct

The booster club bought 8 adult tickets and 32 student tickets.

Check Thirty-two student tickets is 4 times more than the 8 adult tickets. Since the cost of 8 adult tickets, $200, plus the cost of 32 student tickets, $576, equals $776, the guess is correct.

Exercises **Use the guess and check strategy to solve each problem.**

1. **JEWELRY** Jana is making necklaces and bracelets. She puts 8 crystals on each necklace and 3 crystals on each bracelet. She needs to make 20 more necklaces than bracelets. She has 270 crystals. If she uses all the crystals, how many necklaces and bracelets can she make?

2. **GIFT BAGS** The ninth-grade class is filling gift bags for participants in a school fund-raiser. They put 2 raffle tickets in each child's bag and 4 raffle tickets in each adult's bag. They made twice as many adult bags as child bags. If they had 500 raffle tickets, how many child bags and adult bags did they make?

Lesson 8-5

NAME ______________________________ DATE ______________ PERIOD _____

8-5 Practice

Problem-Solving Investigation: Guess and Check

Mixed Problem Solving

For Exercises 1 and 2, solve using the guess and check strategy.

1. **NUMBER THEORY** A number is squared and the result is 676. Find the number.

2. **CRAFTS** Sabrina has 12 spools of ribbon. Each spool has either 3 yards of ribbon, 5 yards of ribbon, or 8 yards of ribbon. If Sabrina has a total of 68 yards of ribbon, how many spools of each length of ribbon does she have?

Use any strategy to solve Exercises 3–7. Some strategies are shown below.

PROBLEM-SOLVING STRATEGIES
• Draw a diagram. • Make a table. • Guess and check.

3. **NUMBERS** Among all pairs of whole numbers with product 66, find the pair with the smallest sum.

4. **SHOPPING** You are buying a jacket that costs \$69.95. If the sales tax rate is 7.75%, would it be more reasonable to expect the sales tax to be about \$4.90 or \$5.60?

5. **STATES** Of the 50 United States, 14 have coastlines on the Atlantic Ocean, 5 have coastlines on the Gulf of Mexico, and one state has coastlines on both. How many states do not have coastlines on either the Atlantic Ocean or the Gulf of Mexico?

6. **TIME** Melissa spent $7\frac{1}{2}$ minutes of the last hour downloading songs from the Internet. What percent of the last hour did she spend downloading songs?

7. **VOLUNTEERING** Greg helps his mother deliver care baskets to hospital patients each Saturday. Last Saturday at noon they had three times as many baskets left to deliver as they had already delivered. If they were delivering a total of 64 baskets that day, how many had they delivered by noon?

8-6 Study Guide and Intervention

Inequalities

A mathematical sentence that contains < or > is called an **inequality.** When used to compare a variable and a number, inequalities can describe a range of values. Some inequalities use the symbols ≤ or ≥. The symbol ≤ is read *is less than or equal to.* The symbol ≥ is read *is greater than or equal to.*

Examples **Write an inequality for each sentence.**

1 **SHOPPING** Shipping is free on orders of more than $100.
Let c = the cost of the order.
$c > 100$

2 **RESTAURANTS** The restaurant seats a maximum of 150 guests.
Let g = the number of guests.
$g \leq 150$

Inequalities can be graphed on a number line. An open or closed circle is used to show where the solutions start, and an arrow pointing either left or right indicates the rest of the solutions. An open circle is used with inequalities having > or <. A closed circle is used with inequalities having ≤ or ≥.

Examples **Graph each inequality on a number line.**

3 $d \leq -2$

Place a closed circle at −2. Then draw a line and an arrow to the left.

–4 –3 –2 –1 0 1 2 3 4

4 $d > -2$

Place an open circle at −2. Then draw a line and an arrow to the right.

Exercises

Write an inequality for each sentence.

1. **FOOD** Our delivery time is guaranteed to be less than 30 minutes.

2. **DRIVING** Your speed must be at least 45 miles per hour on the highway.

Graph each inequality on a number line.

3. $r > 7$

4. $x \leq -1$

NAME ______________________ DATE __________ PERIOD _____

8-6 Practice

Inequalities

Write an inequality for each sentence.

1. **JOBS** Applicants with less than 5 years of experience must take a test.

2. **FOOTBALL** The home team needs more than 6 points to win.

3. **VOTING** The minimum voting age is 18.

4. **GAMES** You must answer at least 10 questions correctly to stay in the game.

5. **DINING** A tip of no less than 10% is considered acceptable.

6. **MONEY** The cost including tax is no more than $75.

For the given value, state whether the inequality is true or false.

7. $9 + b < 16, b = 8$

8. $14 - f > 8, f = 5$

9. $-5t < 24, t = 5$

10. $51 \leq 3m, m = 17$

11. $\frac{z}{5} \leq 7, z = 40$

12. $\frac{-28}{d} > 7, d = -4$

Graph each inequality on a number line.

13. $y > 5$

14. $h < 5$

15. $c \leq 1$

16. $t \geq 2$

17. $x \geq 4$

18. $r < 9$

For Exercises 19 and 20, use the table that shows the literacy rate in several countries.

19. In which country or countries is the literacy rate less than 90%?

20. In which country or countries is the literacy rate at least 88%?

Country	Literacy Rate
Albania	87%
Jamaica	88%
Panama	93%
Senegal	40%

NAME ______________________ DATE ____________ PERIOD _____

8-7 Study Guide and Intervention

Solving Inequalities by Adding or Subtracting

Solving an inequality means finding values for the variable that make the inequality true. You can use the Addition and Subtraction Properties of Inequality to help solve an inequality. When you add or subtract the same number from each side of an inequality, the inequality remains true.

Examples **Solve each inequality. Check your solution. Then graph the solution on a number line.**

1 $9 < r + 5$

$9 < r + 5$	Write the inequality.
$9 - 5 < r + 5 - 5$	Subtract 5 from each side.
$4 < r$ or $r > 4$	Simplify.

Check Solutions to the inequality should be greater than 4. Check this result by replacing r in the original inequality with two different numbers greater than 4. Both replacements should give true statements.

To graph the solution, place an open circle at 4 and draw a line and arrow to the right.

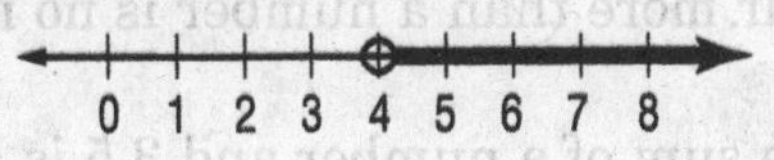

$x - 7 \geq -4$

$x - 7 \geq -4$	Write the inequality.
$x - 7 + 7 \geq -4 + 7$	Add 7 to each side.
$x \geq 3$	Simplify.

Check Replace x in the original inequality with 3 and then with a number greater than 3. The solution is $x \geq 3$.

To graph the solution, place a closed circle at 3 and draw a line and arrow to the right.

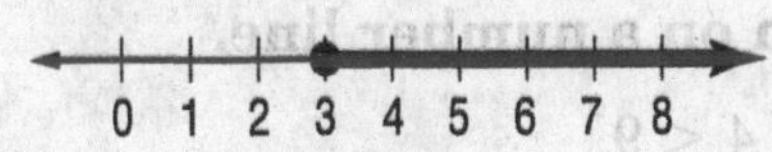

Exercises

Solve each inequality. Check your solution.

1. $t - 4 > 2$ **2.** $b + 5 \leq 9$ **3.** $8 < r - 7$

4. $6 < p - 6$ **5.** $2 > a + 7$ **6.** $4 + m \geq -6$

Solve each inequality and check your solution. Then graph the solution on a number line.

7. $s + 8 < 10$

8. $-11 \leq d - 3$

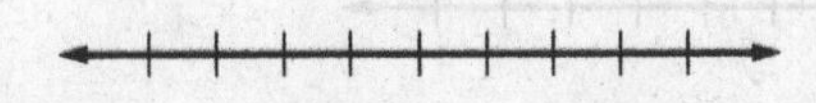

NAME ______________________ DATE __________ PERIOD _____

8-7 Practice

Solving Inequalities by Adding or Subtracting

Solve each inequality. Check your solution.

1. $p + 7 < 9$

2. $t + 6 > -3$

3. $-13 \geq 9 + b$

4. $16 > -11 + k$

5. $3 \geq -2 + y$

6. $25 < n + (-12)$

7. $r - 5 \leq 2$

8. $a - 6 < 13$

9. $j - 8 \leq -12$

10. $-8 > h - 1$

11. $22 > w - (-16)$

12. $-30 \leq d + (-5)$

13. $1 + y \leq 2.4$

14. $b - \frac{3}{4} < 2\frac{1}{2}$

15. $f - 4 \geq 1.4$

Write an inequality and solve each problem.

16. Five less than a number is more than twenty.

17. Four more than a number is no more than twelve.

18. The sum of a number and 3.5 is at least 14.5.

19. The difference of a number and −5 is less than 7.

20. The sum of −12 and a number is at least 6.

21. Eleven less than a number is more than fifteen.

Solve each inequality and check your solution. Then graph the solution on a number line.

22. $n + 4 < 9$

23. $t + 7 > 12$

24. $p + (-5) > -3$

25. $-13 \geq x - 8$

26. $-32 \geq a + (-5)$

27. $3 \leq \frac{1}{2} + m$

28. $4 \geq s - \frac{2}{3}$

29. $-\frac{3}{4} < w - 1$

NAME ______________________________ DATE ____________ PERIOD ______

8-8 Study Guide and Intervention

Solving Inequalities by Multiplying or Dividing

When you multiply or divide each side of an inequality by a positive number, the inequality remains true. However, when you multiply or divide each side of an inequality by a negative number, the direction of the inequality must be reversed for the inequality to remain true.

Example 1 **Solve $\frac{t}{-7} \leq -3$. Check your solution. Then graph the solution on a number line.**

$\frac{t}{-7} \leq -3$	Write the inequality.
$\frac{t}{-7}(-7) \geq -3(-7)$	Multiply each side by −7 and reverse the inequality symbol.
$t \geq 21$	Simplify.

The solution is $t \geq 21$. You can check this solution by replacing t in the original inequality with 21 and a number greater than 21.

To graph the solution, place a closed circle at 21 and draw a line and arrow to the right.

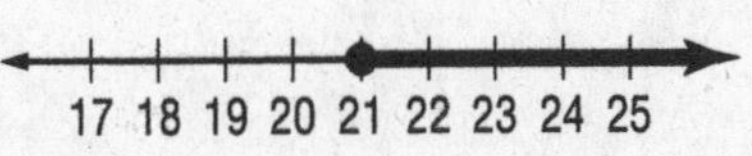

Some inequalities involve more than one operation.

Example 2 **Solve $4x - 5 < 27$. Check your solution.**

$4x - 5 < 27$	Write the inequality.
$4x - 5 + 5 < 27 + 5$	Add 5 to each side.
$4x < 32$	Simplify.
$\frac{4x}{4} < \frac{32}{4}$	Divide each side by 4.
$x < 8$	Simplify.

The solution is $x < 8$. You can check this solution by substituting numbers less than 8 into the original inequality.

Exercises

Solve each inequality and check your solution. Then graph the solution on a number line.

1. $3a > 6$

2. $36 > 4r$

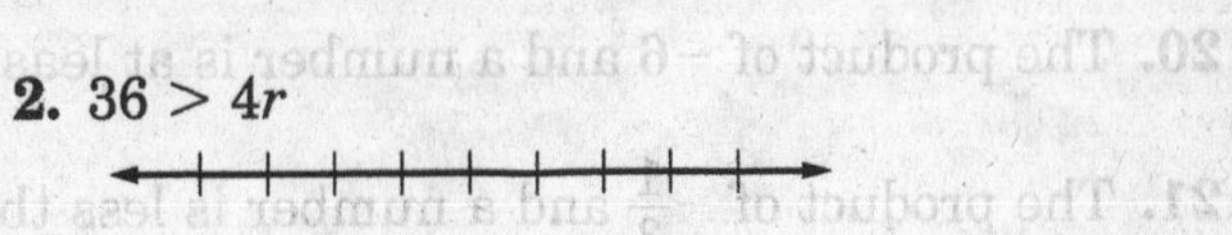

Solve each inequality. Check your solution.

3. $c + 2 \geq -2$

4. $13 > -2y - 3$

5. $\frac{h}{-5} - 6 < -10$

Lesson 8-8

NAME ______________________ DATE __________ PERIOD _____

8-8 Practice

Solving Inequalities by Multiplying or Dividing

Solve each inequality and check your solution. Then graph the solution on a number line.

1. $-8 \leq 4w$

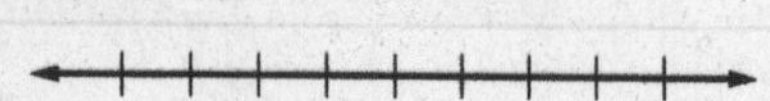

2. $-6a > -78$

3. $-25t \leq 400$

4. $18 > -2g$

5. $\frac{y}{4} \geq 2.4$

6. $\frac{r}{-2} < -2$

7. $-8 > \frac{k}{-0.4}$

8. $\frac{m}{-7} \leq 1.2$

Solve each inequality. Check your solution.

9. $13a \geq -26$

10. $-15 \leq 5b$

11. $-3m > -33$

12. $-8z \leq -24$

13. $\frac{n}{-5} \geq 0.8$

14. $6 > \frac{x}{-7}$

15. $-5 \geq \frac{c}{4.5}$

16. $-19 > \frac{y}{-0.3}$

17. $-\frac{1}{3}x \geq -9$

Write an inequality for each sentence. Then solve the inequality.

18. Five times a number is more than 55.

19. The quotient of a number and 12 is no more than three.

20. The product of -6 and a number is at least 54.

21. The product of $-\frac{1}{3}$ and a number is less than -36.

22. The quotient of a number and 3 is at least -5.

23. A number divided by 4 is more than 16.

NAME ______________________ DATE ____________ PERIOD _____

9-1 Study Guide and Intervention

Sequences

A **sequence** is an ordered list of numbers. Each number in the list is called a **term**. An **arithmetic sequence** is a sequence in which the difference between any two consecutive terms is the same. This difference is called the **common difference**. To find the next number in an arithmetic sequence, add the common difference to the last term.

Example **State whether the sequence −4, −1, 2, 5, 8, . . . is arithmetic. If it is, state the common difference. Write the next three terms of the sequence.**

−4, −1, 2, 5, 8
+3 +3 +3 +3

Notice that $-1 - (-4) = 3$, $2 - (-1) = 3$, and so on. The terms have a common difference of 3, so the sequence is arithmetic.

$8 + 3 = $ **11**, $11 + 3 = $ **14**, $14 + 3 = $ **17**. The next three terms are 11, 14, and 17.

Some sequences are not arithmetic. To extend a sequence like this, look for a pattern in the consecutive differences. Then apply the pattern to the last term of the sequence.

Arithmetic sequences can be described algebraically. Use the table to examine the sequence.

Term Number (n)	1	2	3	4
Term	4	7	10	13

The terms have a common difference of 3. Also, each term is 3 times its term number + 1. An expression that can be used to find the nth term is $3n + 1$. What are the next two terms?

$3(5) + 1 = 15 + 1 = 16$

$3(6) + 1 = 18 + 1 = 19$

Exercises

State whether each sequence is arithmetic. Write *yes* or *no*. If it is, state the common difference and write the next three terms.

1. 0, 3, 6, 9, 12, . . .

2. 3, 6, 12, 24, 48, . . .

3. 6, 11, 16, 21, 26, . . .

4. 0, 1, 3, 6, 10, . . .

5. $\frac{1}{9}, \frac{1}{3}$, 1, 3, 9, . . .

6. 30, 26, 22, 18, 14, . . .

NAME ______________________ DATE ____________ PERIOD ______

9-1 Practice

Sequences

State whether each sequence is arithmetic. Write *yes* or *no*. If it is, state the common difference and write the next three terms.

1. 14, 22, 30, 38, 46, ...

2. 5, 25, 125, 625, 3,125

3. 81, 27, 9, 3, 1, ...

4. 10, 3, −4, −11, −18, ...

5. 3, 7, 12, 18, 25, ...

6. 2, $4\frac{1}{2}$, 7, $9\frac{1}{2}$, 12, ...

Write an expression that can be used to find the *n*th term of each sequence. Then find the next three terms.

7. 3, 6, 9, 12, ...

8. 9, 18, 27, 36, ...

9. $\frac{1}{5}, \frac{2}{5}, \frac{3}{5}, \frac{4}{5}, \ldots$

10. $\frac{3}{7}, \frac{6}{7}, 1\frac{2}{7}, 1\frac{5}{7}, \ldots$

11. 7, 13, 19, 25, ...

12. 1, 6, 11, 16, ...

Write and solve an expresion to find the *n*th term of each arithmetic sequence.

13. 2, 5, 8, 11, ... ; $n = 10$

14. 18, 22, 26, 30, ... ; $n = 14$

15. 12, 6, 0, −6, ... ; $n = 20$

16. 47, 39, 31, 23, ... ; $n = 18$

17. 28, 41, 54, 67, ... ; $n = 100$

18. 74, 91, 108, 125, ... ; $n = 150$

AMUSEMENT PARKS For Exercises 19 and 20, use the following information.

An amusement park offers a daily pass for $42 or a season pass for $210.

Number of Amusement Park Visits	1	2	3	4	5
Total Cost with Daily Passes	$42	$84			
Total Cost with Season Pass	$210	$210			

19. Is the sequence formed by the total cost with daily passes arithmetic? Explain.

20. Is the sequence formed by the total cost with a season pass arithmetic? Explain.

NAME ______________________ DATE __________ PERIOD _____

9-2 Study Guide and Intervention

Functions

A **function** connects an input number, x, to an output number, $f(x)$, by a rule. To find the value of a function for a certain number, substitute the number into the function value in place of x, and simplify.

Example 1 **Find $f(5)$ if $f(x) = 2 + 3x$.**

$f(x) = 2 + 3x$ — Write the function.
$f(5) = 2 + 3(5)$ or 17 — Substitute 5 for x into the function rule and simplify.
So, $f(5) = 17$.

You can organize the input, rule, and output of a function using a function table.

Example 2 **Complete the function table for $f(x) = 2x + 4$.**

Substitute each value of x, or input, into the function rule. Then simplify to find the output.

$f(x) = 2x + 4$
$f(-1) = 2(-1) + 4$ or 2
$f(0) = 2(0) + 4$ or 4
$f(1) = 2(1) + 4$ or 6
$f(2) = 2(2) + 4$ or 8

Input x	Rule $2x + 4$	Output $f(x)$
−1	2(−1) + 4	2
0	2(0) + 4	4
1	2(1) + 4	6
2	2(2) + 4	8

Exercises

Find each function value.

1. $f(1)$ if $f(x) = x + 3$
2. $f(6)$ if $f(x) = 2x$
3. $f(4)$ if $f(x) = 5x - 4$
4. $f(9)$ if $f(x) = -3x + 10$
5. $f(-2)$ if $f(x) = 4x - 1$
6. $f(-5)$ if $f(x) = -2x + 8$

Complete each function table.

7. $f(x) = x - 10$

x	$x - 10$	$f(x)$
−1		
0		
1		
2		

8. $f(x) = 2x + 6$

x	$2x + 6$	$f(x)$
−3		
−1		
2		
4		

9. $f(x) = 2 - 3x$

x	$2 - 3x$	$f(x)$
−2		
0		
3		
4		

NAME ______________________ DATE ____________ PERIOD _____

9-2 Practice

Functions

Find each function value.

1. $f(6)$ if $f(x) = 4x$
2. $f(8)$ if $f(x) = x + 11$
3. $f(3)$ if $f(x) = 2x + 4$
4. $f(5)$ if $f(x) = 3x - 2$
5. $f(-6)$ if $f(x) = 4x + 7$
6. $f(-14)$ if $f(x) = 2x - 3$
7. $f(\frac{2}{9})$ if $f(x) = 3x + \frac{1}{3}$
8. $f(\frac{3}{4})$ if $f(x) = 2x - \frac{1}{4}$
9. $f(\frac{4}{5})$ if $f(x) = 4x - \frac{1}{5}$

Complete each function table. Then state the domain and range of the function.

10. $f(x) = 5x - 4$

x	$5x - 4$	$f(x)$
−4		
−1		
3		
6		

11. $f(x) = 2 - 3x$

x	$2 - 3x$	$f(x)$
−3		
0		
2		
5		

12. $f(x) = 6 + 2x$

x	$6 + 2x$	$f(x)$
−3		
−1		
1		
4		

13. $f(x) = x - 7$

x	$x - 7$	$f(x)$
−3		
−2		
5		
10		

14. $f(x) = 9x$

x	$9x$	$f(x)$
−6		
−4		
1		
3		

15. $f(x) = 3x + 5$

x	$3x + 5$	$f(x)$
−5		
−1		
2		
6		

16. **JACKETS** The school baseball team wants to have each player's name imprinted on the player's jacket. The cost is \$75 plus \$8.50 for each name. Write a function to represent the cost c for n names. What is the cost to have names imprinted on 25 jackets?

17. **LEMONADE** Gene sold 10 glasses of lemonade while setting up his lemonade stand. After opening, he sold an average of 20 glasses each hour. Write a function to represent the approximate number of glasses g sold after h hours. About when did he sell the 100th glass of lemonade?

NAME ______________________ DATE __________ PERIOD _____

9-3 Study Guide and Intervention

Representing Linear Functions

A function in which the graph of the solutions forms a line is called a **linear function.** A linear function can be represented by an equation, a table, a set of ordered pairs, or a graph.

Example **Graph $y = x - 2$.**

Step 1 Choose some values for x. Use these values to make a function table.

x	$x - 2$	y	(x, y)
0	0 − 2	−2	(0, −2)
1	1 − 2	−1	(1, −1)
2	2 − 2	0	(2, 0)
3	3 − 2	1	(3, 1)

Step 2 Graph each ordered pair on a coordinate plane. Draw a line that passes through the points. The line is the graph of the linear function.

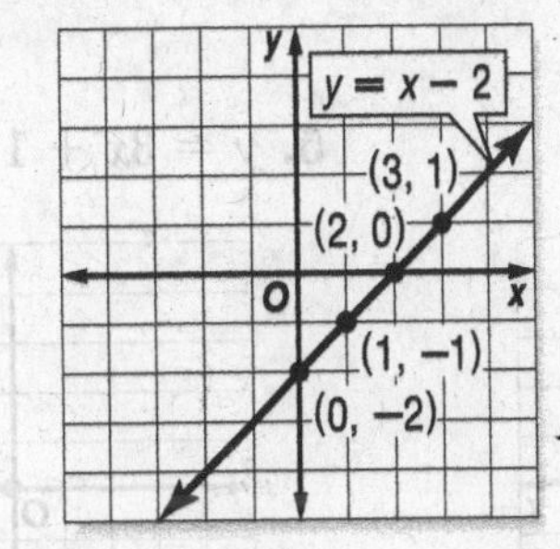

Exercises

Complete the function table. Then graph the function.

1. $y = x + 3$

x	$x + 3$	y	(x, y)
−2			
0			
1			
2			

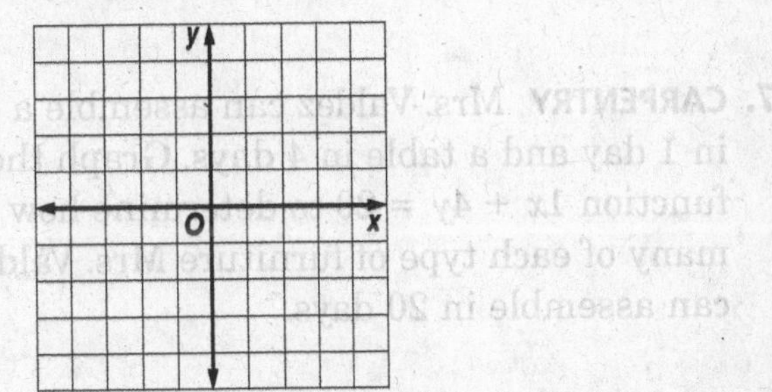

Graph each function.

2. $y = 3x + 2$

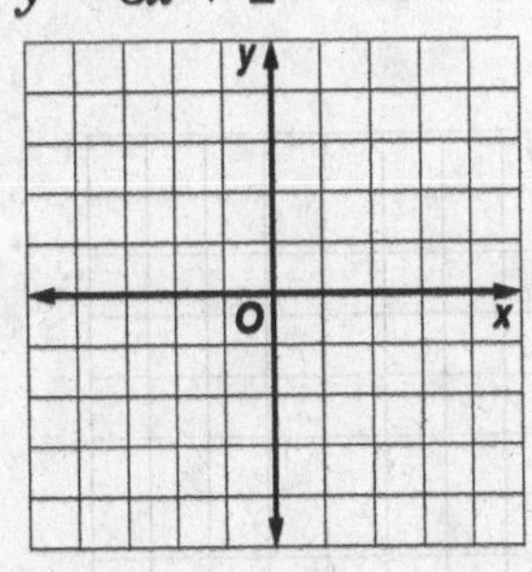

3. $y = 2 - x$

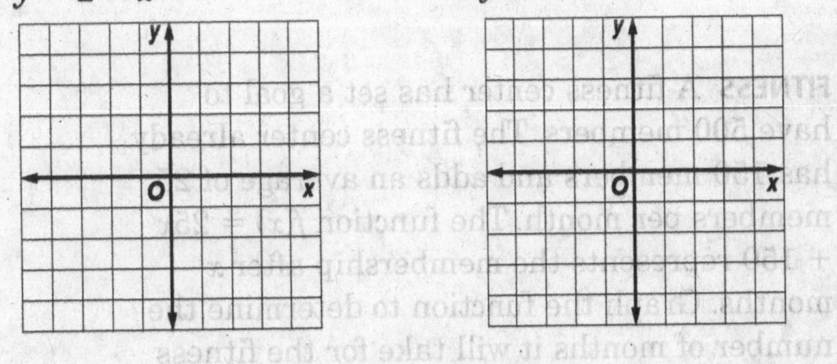

4. $y = 3x - 1$

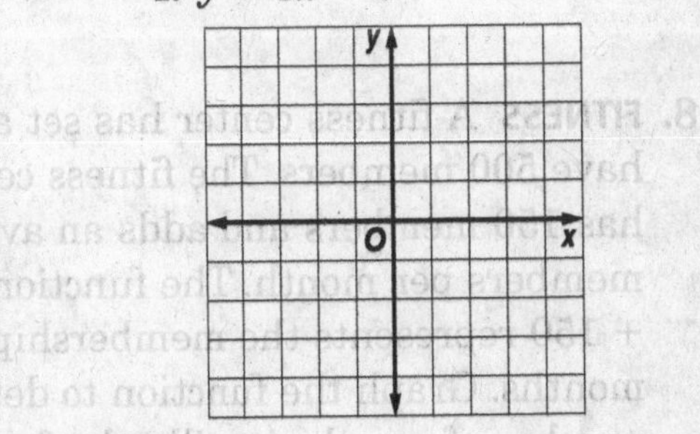

Lesson 9-3

9-3 Practice

Representing Linear Functions

Graph each function.

1. $y = 2x$

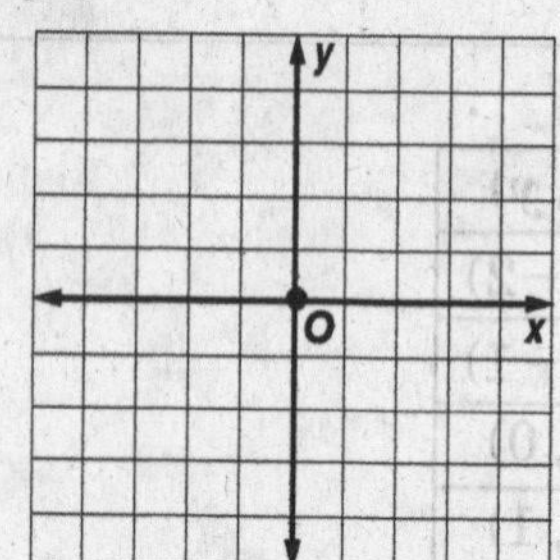

2. $y = -4x$

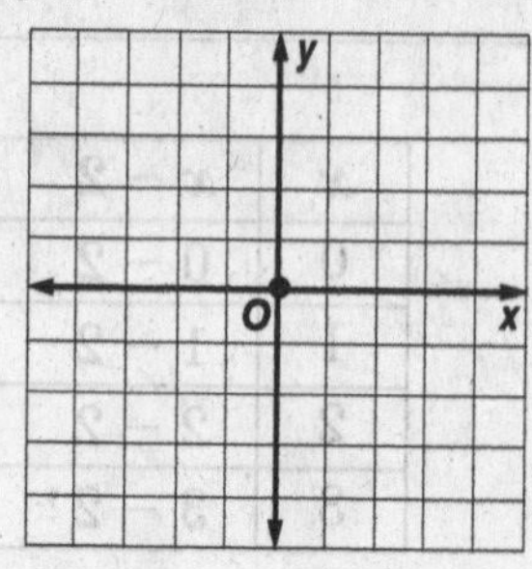

3. $y = x - 4$

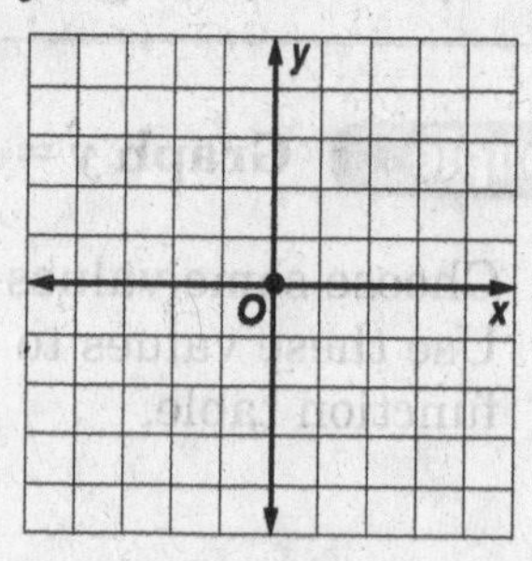

4. $y = x + 3$

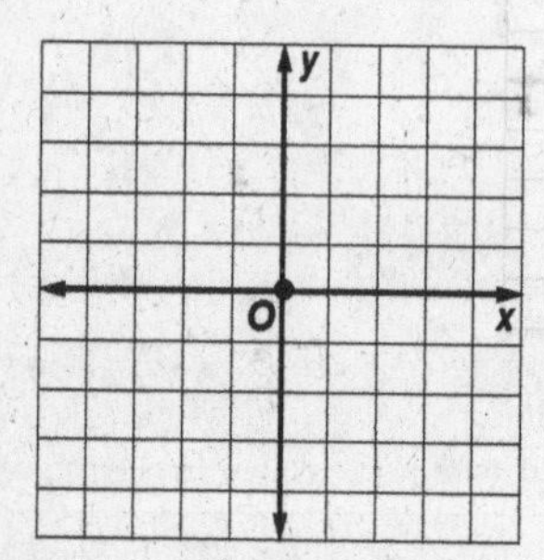

5. $y = 3x + 1$

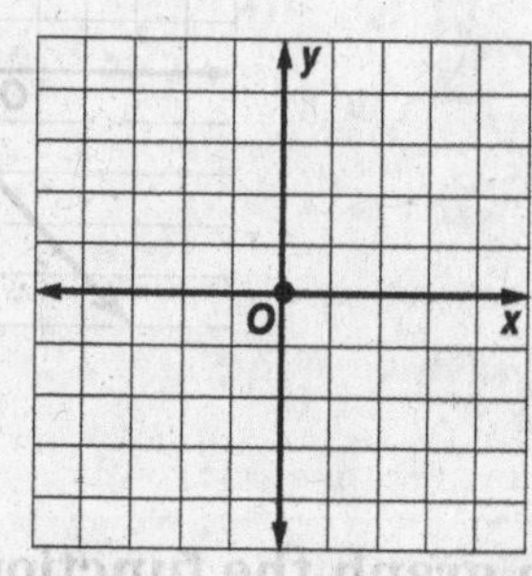

6. $y = \frac{1}{4}x + 2$

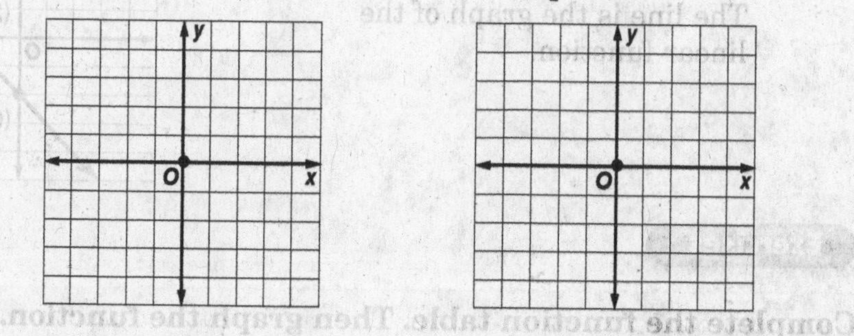

7. **CARPENTRY** Mrs. Valdez can assemble a chair in 1 day and a table in 4 days. Graph the function $1x + 4y = 20$ to determine how many of each type of furniture Mrs. Valdez can assemble in 20 days.

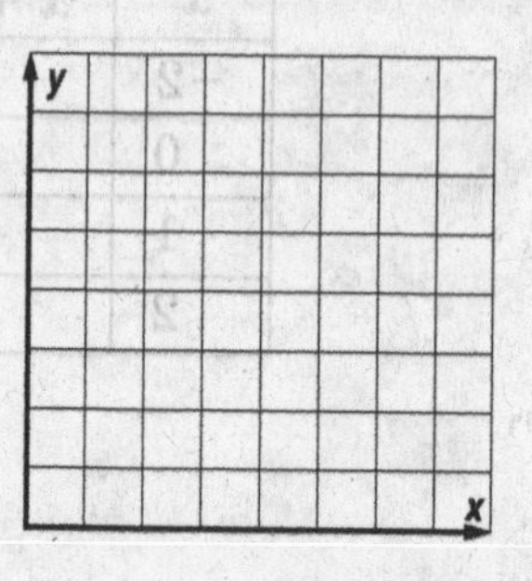

8. **FITNESS** A fitness center has set a goal to have 500 members. The fitness center already has 150 members and adds an average of 25 members per month. The function $f(x) = 25x + 150$ represents the membership after x months. Graph the function to determine the number of months it will take for the fitness center to reach its membership goal.

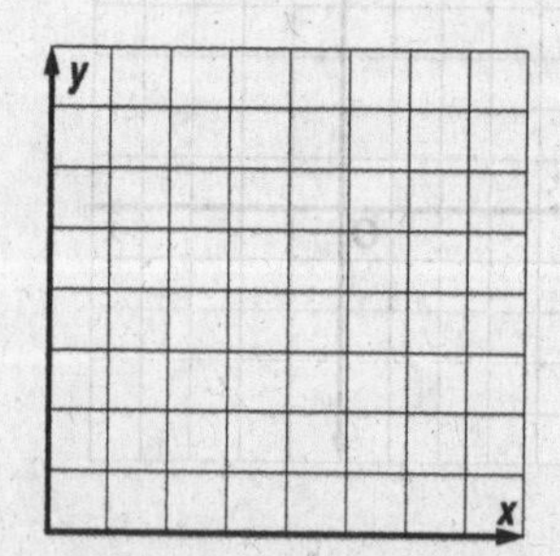

NAME ______________________ DATE __________ PERIOD ______

9-4 Study Guide and Intervention

Slope

The slope m of a line passing through points (x_1, y_1) and (x_2, y_2) is the ratio of the difference in the y-coordinates to the corresponding difference in the x-coordinates. As an equation, the slope is given by

$$m = \frac{y_2 - y_1}{x_2 - x_1}, \text{ where } x_1 \neq x_2.$$

Example 1 **Find the slope of the line that passes through $A(-1, -1)$ and $B(2, 3)$.**

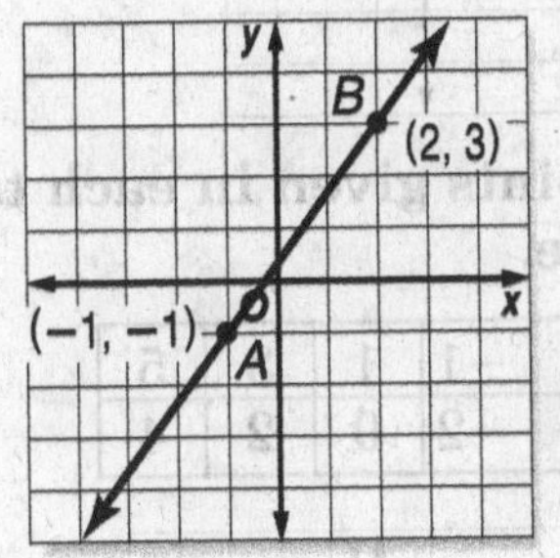

$m = \frac{y_2 - y_1}{x_2 - x_1}$ Definition of slope

$m = \frac{3 - (-1)}{2 - (-1)}$ $(x_1, y_1) = (-1, -1)$, $(x_2, y_2) = (2, 3)$

$m = \frac{4}{3}$ Simplify.

Check When going from left to right, the graph of the line slants upward. This is correct for a positive slope.

Example 2 **Find the slope of the line that passes through $C(1, 4)$ and $D(3, -2)$.**

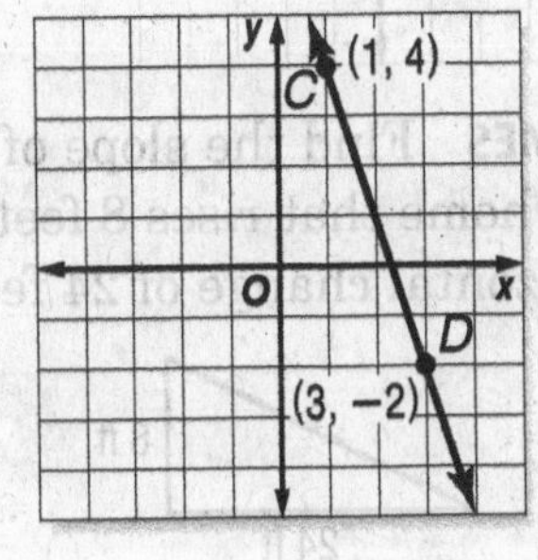

$m = \frac{y_2 - y_1}{x_2 - x_1}$ Definition of slope

$m = \frac{-2 - 4}{3 - 1}$ $(x_1, y_1) = (1, 4)$, $(x_2, y_2) = (3, -2)$

$m = \frac{-6}{2}$ or -3 Simplify.

Check When going from left to right, the graph of the line slants downward. This is correct for a negative slope.

Exercises

Find the slope of the line that passes through each pair of points.

1. $A(0, 1), B(3, 4)$

2. $C(1, -2), D(3, 2)$

3. $E(4, -4), F(2, 2)$

4. $G(3, 1), H(6, 3)$

5. $I(4, 3), J(2, 4)$

6. $K(-4, 4), L(5, 4)$

Lesson 9-4

NAME ______________________________ DATE ____________ PERIOD _____

9-4 Practice

Slope

Find the slope of each line.

1.

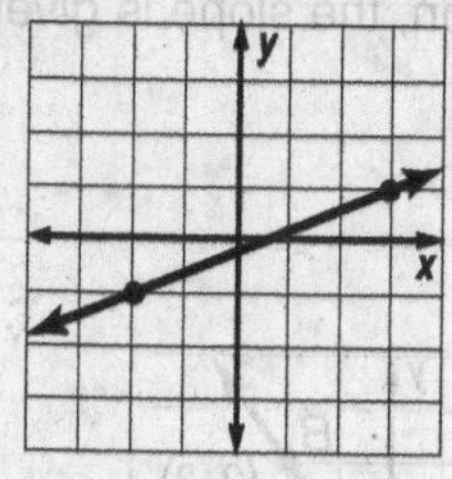

2.

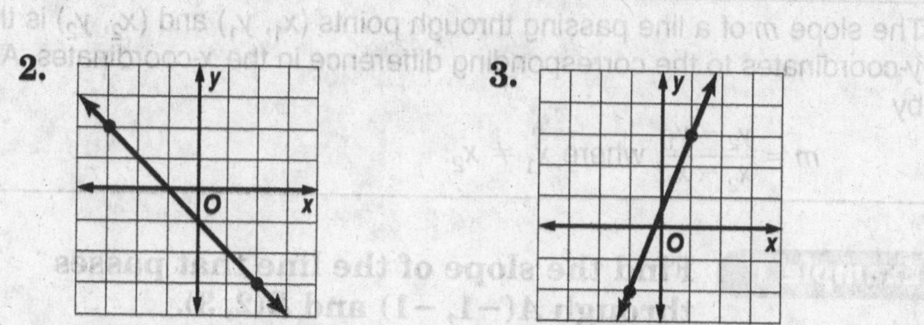

3.

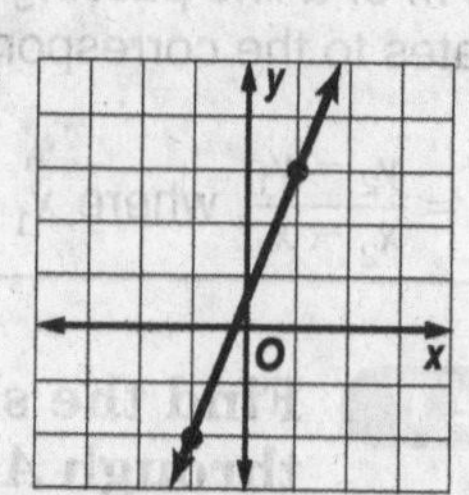

The points given in each table lie on a line. Find the slope of the line. Then graph the line.

4.

x	−1	1	3	5
y	−2	0	2	4

5.

x	−2	3	8	13
y	−2	−1	0	1

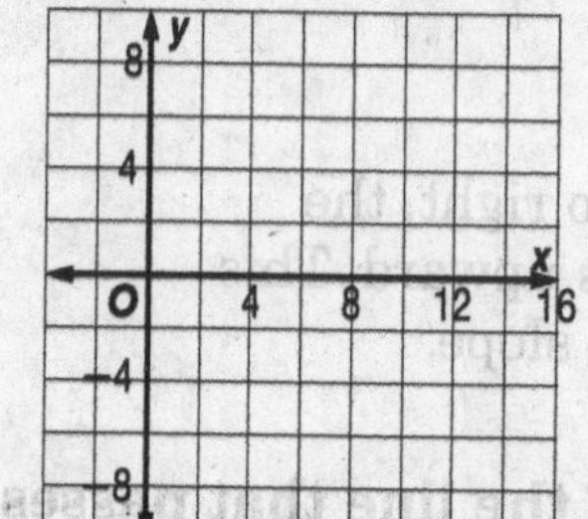

6.

x	−1	2	5	8
y	3	−1	−5	−9

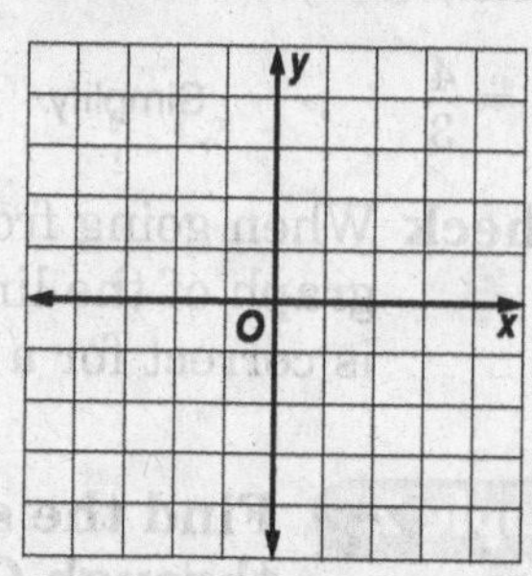

7. **HOMES** Find the slope of the roof of a home that rises 8 feet for every horizontal change of 24 feet.

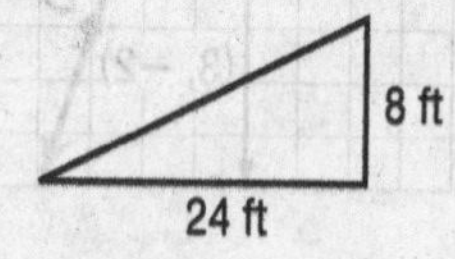

8. **MOUNTAINS** Find the slope of a mountain that descends 100 meters for every horizontal distance of 1,000 meters.

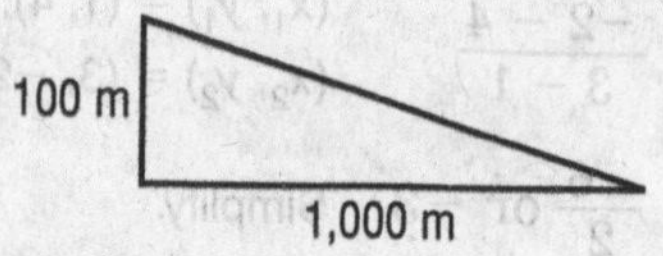

Find the slope of the line that passes through each pair of points.

9. $A(1, 3), B(4, 7)$

10. $C(3, 5), D(2, 6)$

11. $E(4, 0), F(5, 5)$

12. $P(-2, -5), R(2, 3)$

13. $S(-7, 4), T(5, 2)$

14. $V(9, -1), W(7, 6)$

SNOWFALL For Exercises 15–17, use the graph at the right. It shows the depth in feet of snow after each two-hour period during a snowstorm.

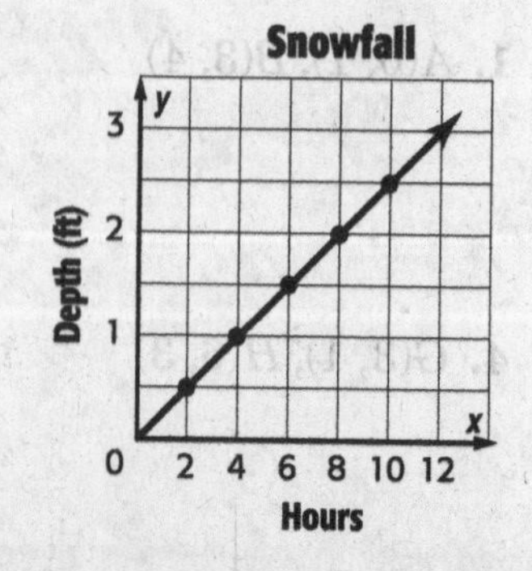

15. Find the slope of the line.

16. Does the graph show a constant rate of change? Explain.

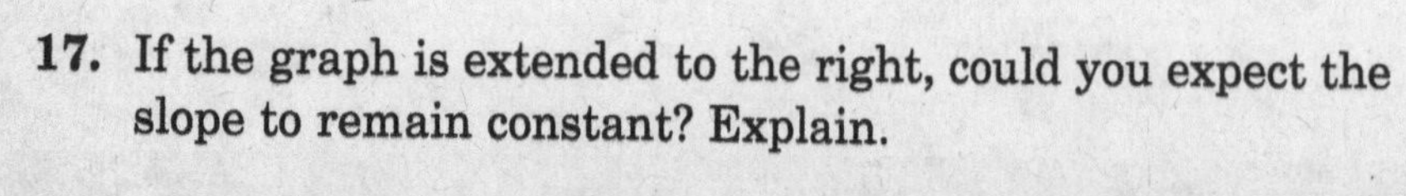

17. If the graph is extended to the right, could you expect the slope to remain constant? Explain.

NAME ______________________ DATE ____________ PERIOD _____

9-5 Study Guide and Intervention

Direct Variation

When two variable quantities have a constant ratio, their relationship is called a **direct variation.**

Example 1 **The distance that a bicycle travels varies directly with the number of rotations that its tires make. Determine the distance that the bicycle travels for each rotation.**

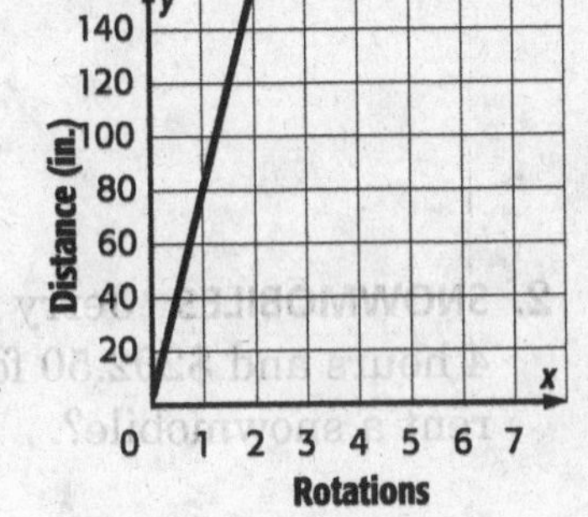

Since the graph of the data forms a line, the rate of change is constant. Use the graph to find the constant ratio.

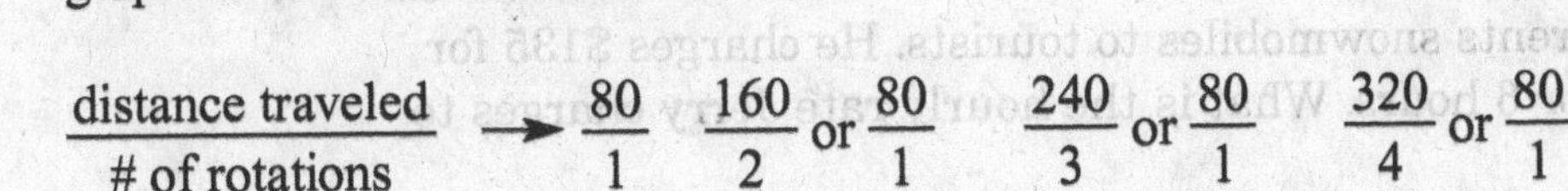

$\dfrac{\text{distance traveled}}{\text{\# of rotations}} \rightarrow \dfrac{80}{1}$ $\dfrac{160}{2}$ or $\dfrac{80}{1}$ $\dfrac{240}{3}$ or $\dfrac{80}{1}$ $\dfrac{320}{4}$ or $\dfrac{80}{1}$

The bicycle travels 80 inches for each rotation of the tires.

Example 2 **The number of trading cards varies directly as the number of packages. If there are 84 cards in 7 packages, how many cards are in 12 packages?**

Let x = the number of packages, y = the total number of cards, and z = the number of cards in each package.

$y = zx$	direct variation
$84 = z(7)$	$y = 84, x = 7$
$12 = z$	Simplify.
$y = 12x$	Substitute for $z = 12$.

Use the equation to find y when $x = 12$.

$y = 12x$	
$y = 12(12)$	$x = 12$
$y = 144$	Multiply.

There are 144 cards in 12 packages.

Exercises

For Exercises Write an expression and solve the given situation.

1. **TICKETS** Four friends bought movie tickets for \$41. The next day seven friends bought movie tickets for \$71.75. What is the price of one ticket?

2. **JOBS** Rick earns \$24.75 in three hours. If the amount that earns varies directly with the number of hours, how much would he earn in 20 hours?

3. **BAKING** A bread recipe calls for $2\frac{1}{2}$ cups of flour for 16 servings, and $3\frac{1}{8}$ cups of flour for 20 servings. How much flour is required to make bread to serve 12?

Lesson 9-5

NAME ________________________ DATE ____________ PERIOD _____

9-5 Practice

Direct Variation

1. **ADVERTISING** The number of vehicles a dealership sells is directly proportional to the money spent on advertising. How many vehicles does a dealership sell for each $1,000 spent on advertising?

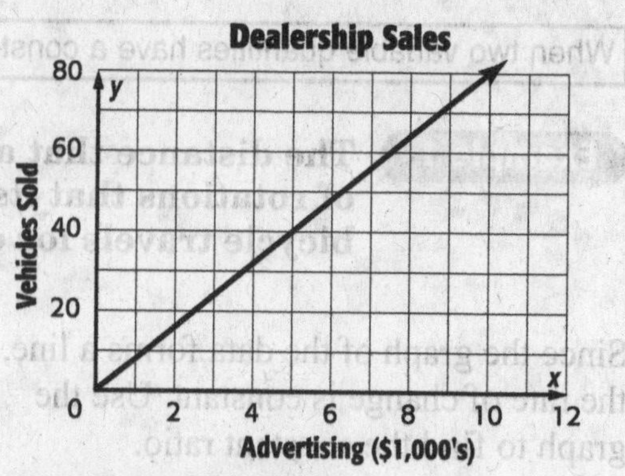

2. **SNOWMOBILES** Jerry rents snowmobiles to tourists. He charges $135 for 4 hours and $202.50 for 6 hours. What is the hourly rate Jerry charges to rent a snowmobile?

3. **SOLAR ENERGY** The power absorbed by a solar panel varies directly with its area. If an 8 square meter panel absorbs 8,160 watts of power, how much power does a 12 square meter solar panel absorb?

4. **INSECT CONTROL** Mr. Malone used 40 pounds of insecticide to cover 1,760 square feet of lawn and 60 pounds to cover an additional 2,640 square feet. How many pounds of insecticide would Mr. Malone need to cover his whole lawn of 4,480 square feet?

Determine whether each linear function is a direct variation. If so, state the constant of variation.

5.

Volume x	2	4	6	8
Mass y	10	20	30	40

6.

Gallons x	5	10	15	20
Miles y	95	190	285	380

7.

Time x	8	9	10	11
Temp y	68	71	74	77

8.

Age x	3	6	9	12
Height y	28	40	52	64

ALGEBRA If y varies directly with x, write an equation for the direct variation. Then find each value.

9. If $y = -5$ when $x = 2$, find y when $x = 8$.

10. Find y when $x = 1$, if $y = 3$ when $x = 2$.

11. If $y = -7$ when $x = -21$, what is the value of x when $y = 9$?

12. Find x when $y = 18$, if $y = 5$ when $x = 4$.

NAME ______________________________ DATE ____________ PERIOD _____

9-6 Study Guide and Intervention

Slope-Intercept Form

Linear equations are often written in the form $y = mx + b$. This is called the **slope-intercept form.** When an equation is written in this form, m is the slope and b is the y-intercept.

Example 1 **State the slope and y-intercept of the graph of $y = x - 3$.**

$y = x - 3$ Write the original equation.

$y = 1x + (-3)$ Write the equation in the form $y = mx + b$.

$\uparrow \quad \uparrow$

$y = mx + b$ $m = 1, b = -3$

The slope of the graph is 1, and the y-intercept is -3.

You can use the slope-intercept form of an equation to graph the equation.

Example 2 **Graph $y = 2x + 1$ using the slope and y-intercept.**

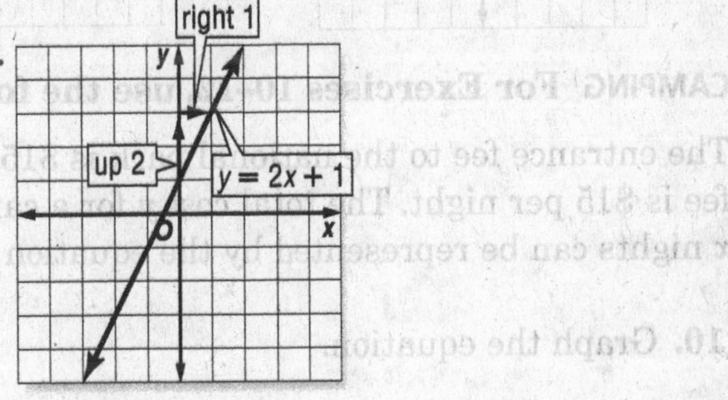

Step 1 Find the slope and y-intercept.
$y = 2x + 1$ slope $= 2$, y-intercept $= 1$.

Step 2 Graph the y-intercept 1.

Step 3 Write the slope 2 as $\frac{2}{1}$. Use it to locate a second point on the line.

$m = \frac{2}{1}$ ← change in y : up 2 units; ← change in x : right 1 unit

Step 4 Draw a line through the two points.

Exercises

State the slope and y-intercept of the graph of each equation.

1. $y = x + 1$

2. $y = 2x - 4$

3. $y = \frac{1}{2}x - 1$

Graph each equation using the slope and y-intercept.

4. $y = 2x + 2$

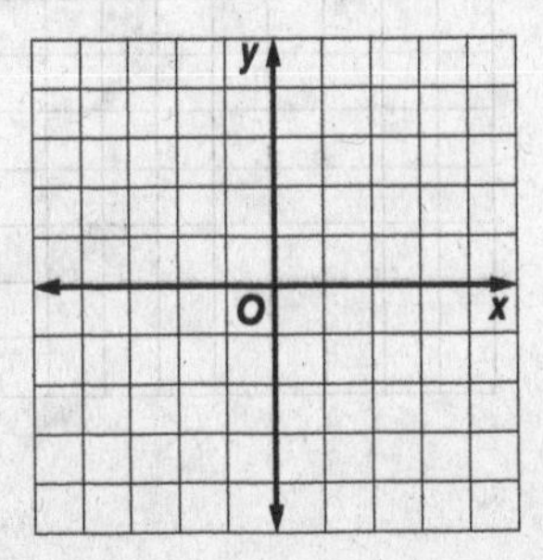

5. $y = x - 1$

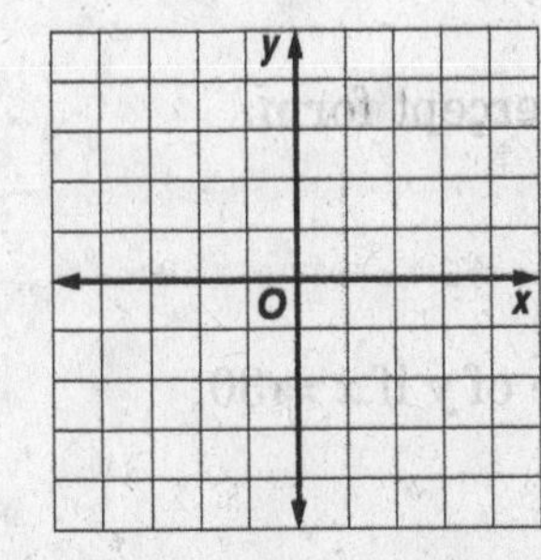

6. $y = \frac{1}{2}x + 2$

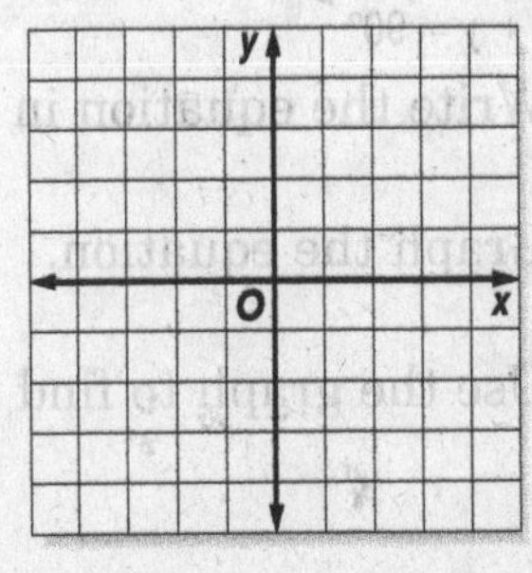

9-6 Practice

Slope-Intercept Form

State the slope and the y-intercept for the graph of each equation.

1. $y = 4x + 1$
2. $y = -3x + 5$
3. $-x + y = 4$
4. $y = -\frac{5}{6}x - 3$
5. $y + 3x = -7$
6. $y = \frac{1}{5}x + 2$

Graph each equation using the slope and the y-intercept.

7. $y = -2x + 2$
8. $y + x = -3$
9. $1 = y - \frac{2}{3}x$

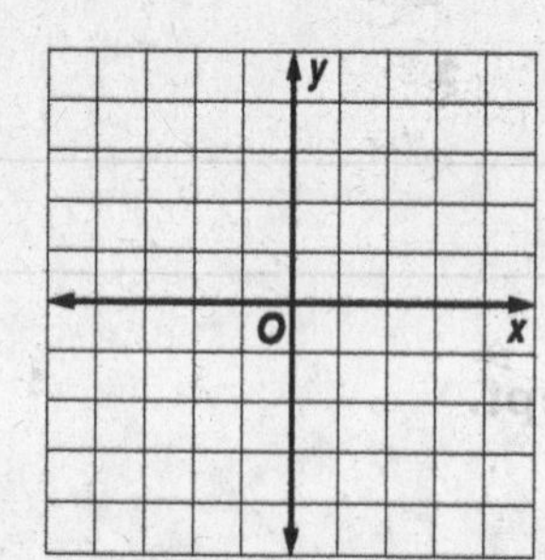

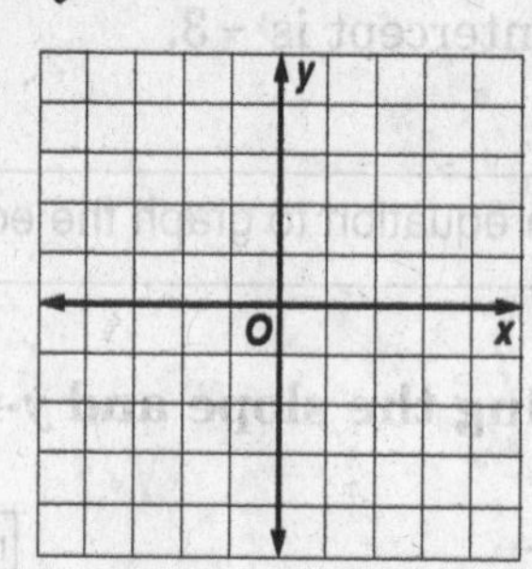

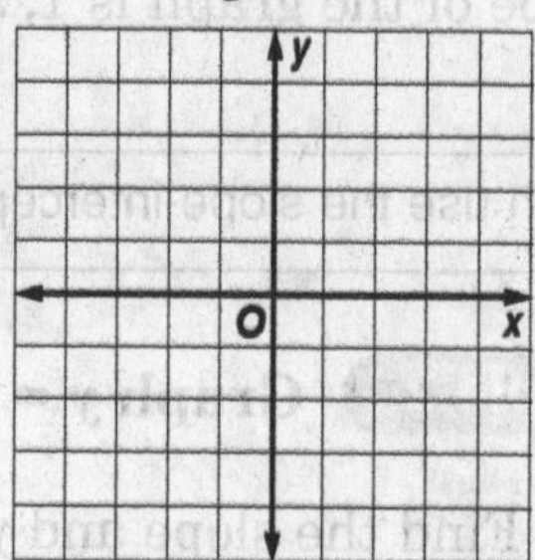

CAMPING For Exercises 10–12, use the following information.

The entrance fee to the national park is \$15. A campsite fee is \$15 per night. The total cost y for a camping trip for x nights can be represented by the equation $y = 15x + 15$.

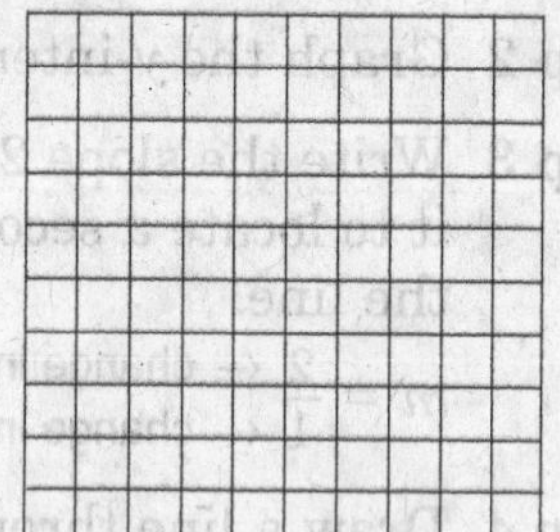

10. Graph the equation.

11. Use the graph to find the total cost for 4 nights.

12. What do the slope and the y-intercept represent?

GEOMETRY For Exercises 13–15, use the diagram shown.

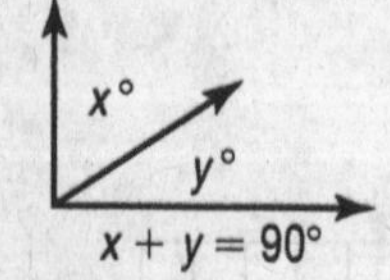

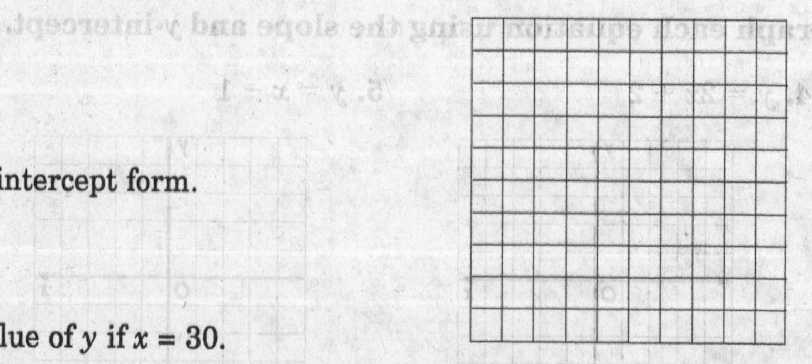

13. Write the equation in slope-intercept form.

14. Graph the equation.

15. Use the graph to find the value of y if $x = 30$.

NAME ______________________ DATE ____________ PERIOD _____

9-7 Study Guide and Intervention

Systems of Equations

Together, the equations $y = x + 3$ and $y = 2x + 1$ are called a **system of equations.** There are two equations and two different unknowns, x and y. The solution of a system of equations is an ordered pair that satisfies each equation.

Example **Two students took a mathematics quiz. Their combined score was 172. Student A's score was 18 points higher than Student B's score. Write a system of equations to represent this situation.**

Step 1 *Choose variables to represent each student.*

Let a = Student A's score and let b = Student B's score

Step 2 *Select the different scenarios to be represented by equations with those variables.*

Their combined score was 172 and Student B's score was 18 points higher than Student A's.

Step 3 *Write a system of equations to represent the situation.*

$a + b = 172$ and $a = b + 18$

Exercises

Write a system of equations to represent each situation.

1. Bianca has a total of 356 CDs and DVDs. She has 24 more CDs than DVDs.

2. Tina wants to spend thirty minutes total studying for math and history today. She also wants to spend five more minutes studying for history than for math.

3. Sixty-two people went to the movies. The price for adults was $6.00 and the price for students was $4.00. The total cost for the group was $290.00.

Lesson 9-7

NAME ______________________________ DATE __________ PERIOD _____

9-7 Practice

Systems of Equations

Solve each system of equations by graphing.

1. $y = 3x + 4$
$y = -x - 4$

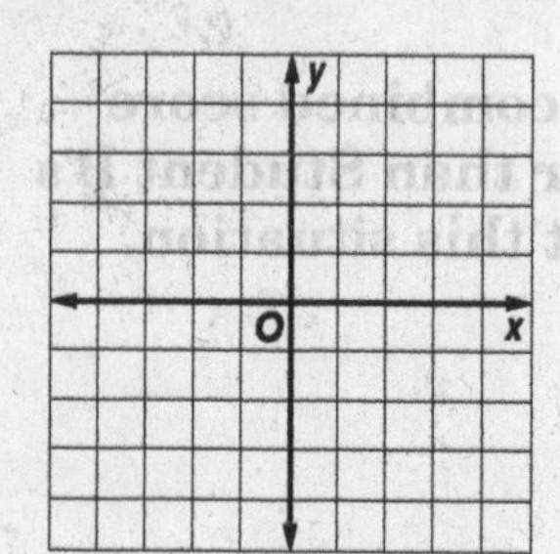

2. $y = 10 + 6x$
$y = 6x$

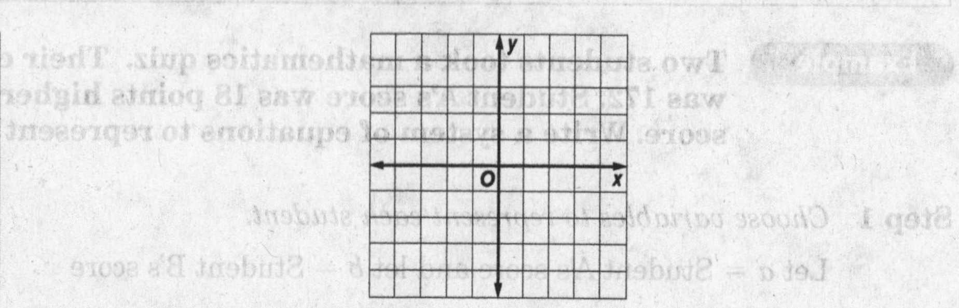

Write and solve a system of equations to represent each situation.

3. Reggie and Miguel scored a total of 54 points in the basketball game. Miguel scored four more points than Reggie.

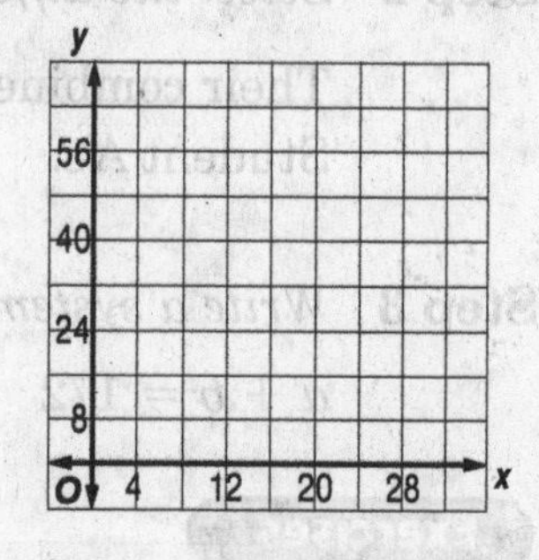

4. Morgan is 15 years younger than Mrs. Santos. Their combined age is 44.

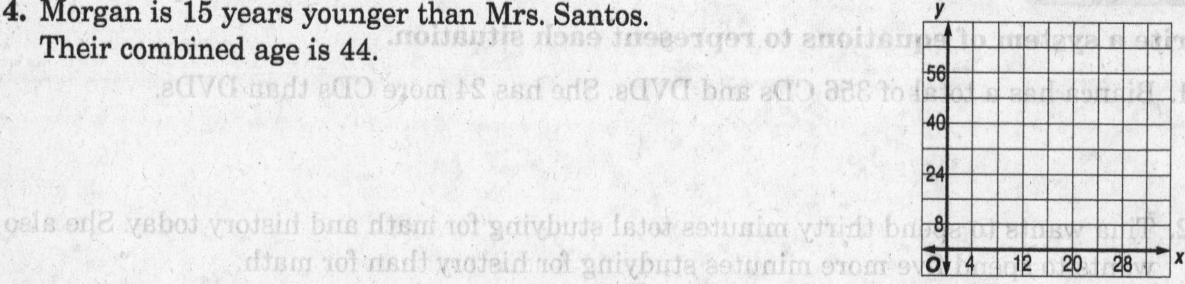

5. The total number of cats and dogs at the shelter is 125. There are 5 more cats than dogs.

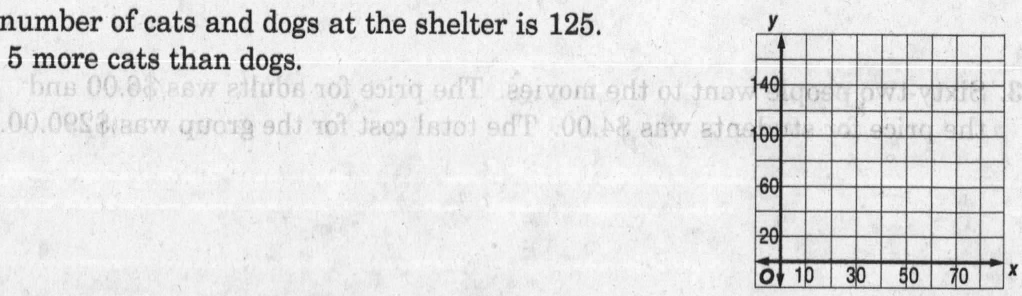

6. Jenny won the ping-pong championship eight more times than Gerardo. They have won a combined total of 32 championships.

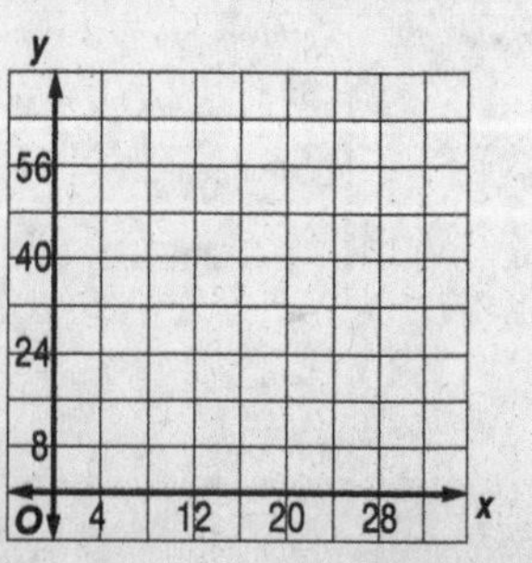

9-8 Study Guide and Intervention

Problem-Solving Investigation: Use a Graph

Example 1 **The graph shows the results of a survey of teachers' ages and grade levels taught at school. Do the oldest teachers teach the highest grade level?**

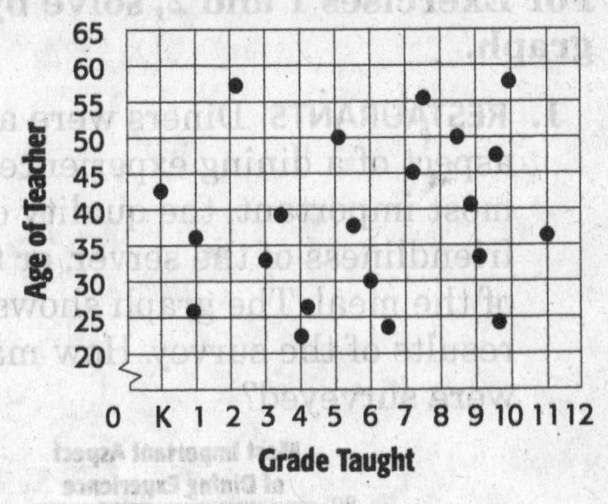

Study the graph. The teachers who are oldest are plotted towards the top of the graph. The teachers who teach the highest grade levels are plotted towards the right of the graph. The graph shows that the points towards the top of the graph are spread out from left to right randomly. The graph shows that the oldest teachers teach all grade levels, not just the highest grade levels.

Example 2 **The graph shows the results of a survey of students' favorite sports. How many students were surveyed?**

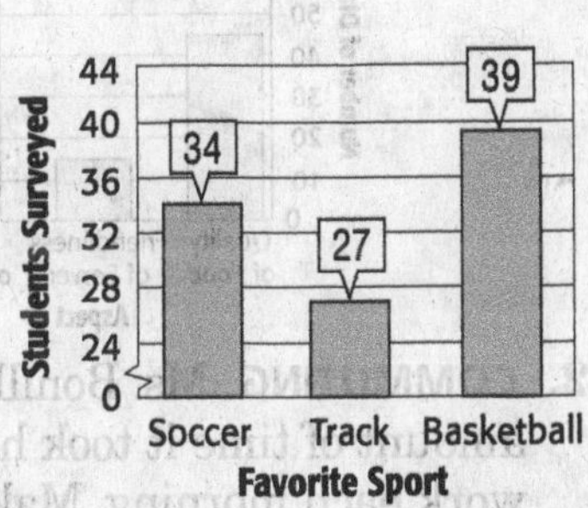

Study the graph. Each bar on the graph represents the number of students who voted for that sport as their favorite. In order to find the number of students surveyed, add the amount from each sport.

soccer + track + basketball = total students surveyed	Write an equation.
34 + 27 + 39 = total students surveyed	Substitute.
100 = total students surveyed	Add.

There were 100 students surveyed.

Exercises

Use the graph at the right. Each point on the graph represents one person in a group that is training for a long-distance bicycle ride. The point shows the number of miles that person cycles each day and the number of weeks that person has been in training.

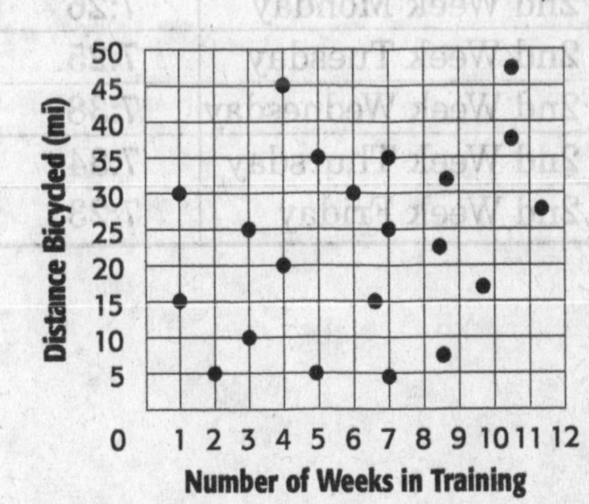

1. What is the highest number of miles bicycled each day by any person in the group? How many weeks was this person in training?

2. Does the number of miles bicycled each day increase as the number of weeks in training increases?

9-8 Practice

Problem-Solving Investigation: Use a Graph

Mixed Problem Solving

For Exercises 1 and 2, solve by using a graph.

1. **RESTAURANTS** Diners were asked which aspect of a dining experience was the most important, the quality of food, the friendliness of the server, or the cost of the meal. The graph shows the results of the survey. How many diners were surveyed?

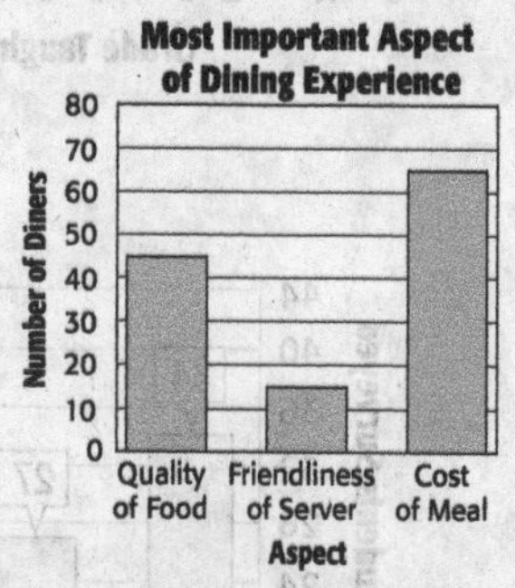

2. **COMMUTING** Ms. Bonilla recorded the amount of time it took her to drive to work each morning. Make a graph of the data in the table. Does the earliest departure time have the least travel time?

Day	Departure Time (A.M.)	Travel Time (min)
1st Week Monday	7:21	17
1st Week Tuesday	7:38	26
1st Week Wednesday	7:32	22
1st Week Thursday	7:20	15
1st Week Friday	7:35	22
2nd Week Monday	7:26	20
2nd Week Tuesday	7:25	18
2nd Week Wednesday	7:38	24
2nd Week Thursday	7:34	21
2nd Week Friday	7:23	17

Use any strategy to solve Exercises 3–5. Some strategies are shown below.

PROBLEM-SOLVING STRATEGIES
- Look for a pattern.
- Use a graph.
- Use logical reasoning.

3. **FLORIST** Ms. Parker charges $29.95 for a bouquet of one dozen roses. Last year, she paid her supplier $4.50 per dozen roses. This year, she paid $3.25 more per dozen. How much less profit did she make this year on 20 dozen bouquets?

4. **TOUR BUS** One line in the graph shows the cost of operating a tour bus. The other line shows the amount of money received from the passengers. How many passengers must ride the tour bus to make a profit?

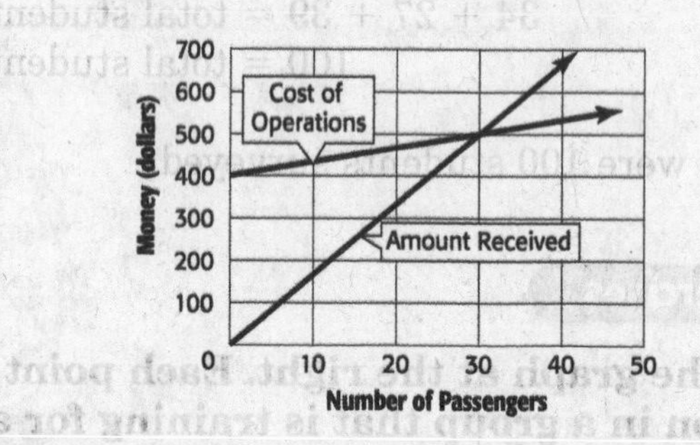

5. **TOWN MEETING** The Waynesville auditorium seats 375 people. In a survey of 50 residents, 6 stated that they plan to attend the next town hall meeting. If the town has 4,200 residents, how many would you expect to attend? Is the auditorium large enough?

9-9 Study Guide and Intervention

Scatter Plots

When you graph two sets of data as ordered pairs, you make a **scatter plot**. The pattern of the data points determines the relationship between the two sets of data.

- Data points that go generally upward show a *positive* relationship.
- Data points that go generally downward show a *negative* relationship.
- Data points with no clear pattern show *no* relationship between the data sets.

Examples **Explain whether the scatter plot of the data shows a *positive*, *negative*, or *no* relationship.**

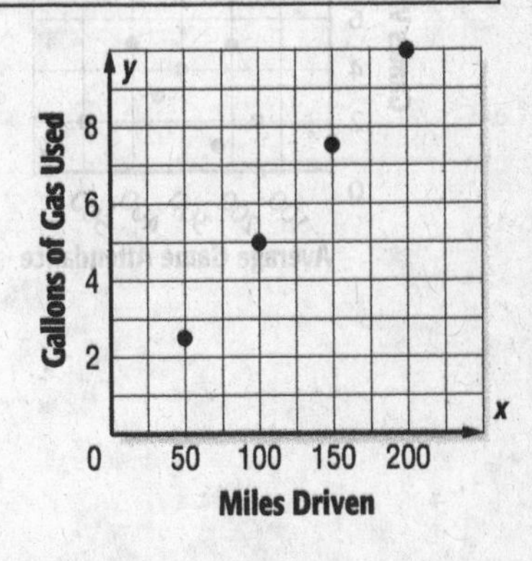

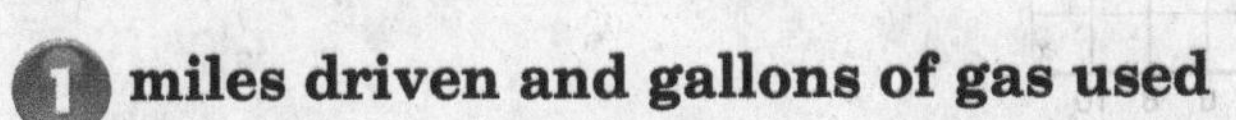

1 miles driven and gallons of gas used

As the number of miles driven increases, the amount of gas used increases. Therefore, the scatter plot will show a positive relationship.

2 number of minutes a candle burns and a candle's height

As the number of minutes increases, the height of the candle will decrease. Therefore, the scatter plot will show a negative relationship.

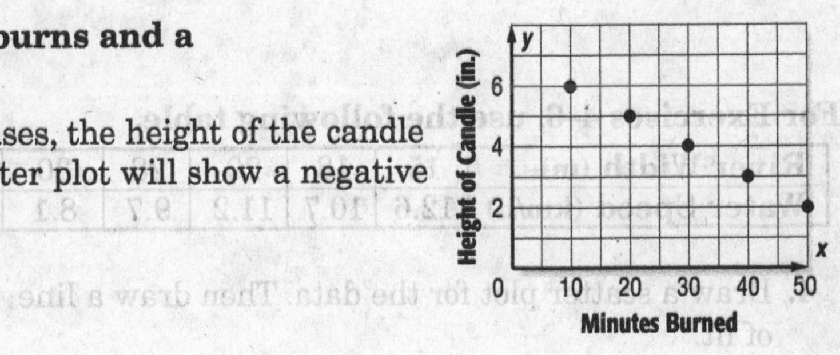

Exercises

Explain whether the scatter plot of the data for the following shows a *positive*, *negative*, or *no* relationship.

1. a student's age and the student's grade level in school

2. number of words written and amount of ink remaining in a pen

3. square feet of floor space and the cost of carpet for the entire floor

4. a person's height and the number of siblings the person has

5. length of time for a shower and the amount of hot water remaining

6. number of sides of a polygon and the area of the polygon

NAME ______________________ DATE ____________ PERIOD _____

9-9 Practice

Scatter Plots

Explain whether the scatter plot of the data for each of the following shows a *positive, negative,* or *no* relationship.

1\.

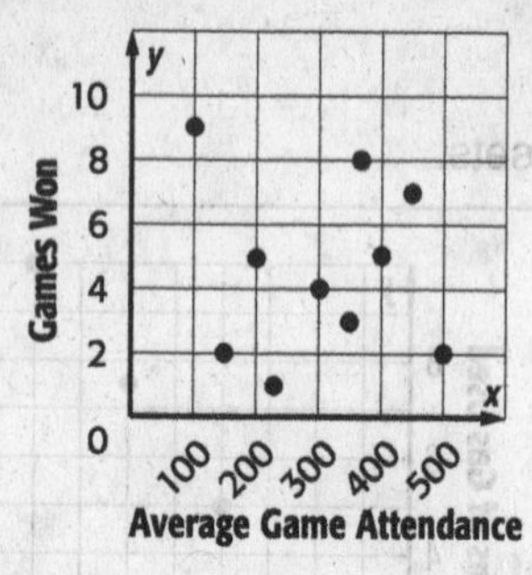

2\.

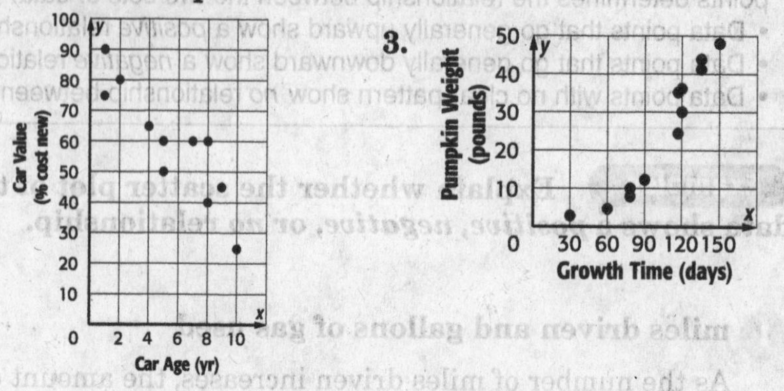

3\.

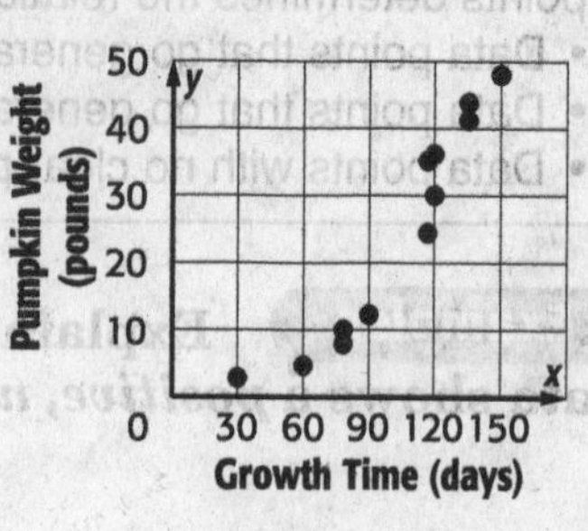

For Exercises 4–6, use the following table.

River Width (m)	15	18	20	28	30	32	38	40	42	45
Water Speed (km/h)	12.6	10.7	11.2	9.7	8.1	8.7	6.9	5.4	3.9	4.1

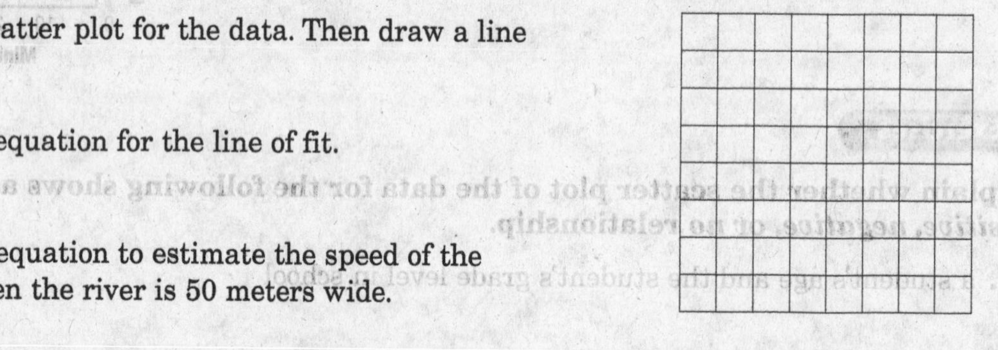

4. Draw a scatter plot for the data. Then draw a line of fit.

5. Write an equation for the line of fit.

6. Use your equation to estimate the speed of the water when the river is 50 meters wide.

Explain whether a scatter plot of the data for each of the following would show a *positive, negative,* or *no* relationship.

7. daily attendance at an outdoor carnival and the number of hours of rain

8. number of diagonals of a polygon and the number of sides of a polygon

NAME ______________________________ DATE ______________ PERIOD _____

10-1 Study Guide and Intervention

Linear and Nonlinear Functions

Linear functions, which have graphs that are straight lines, represent constant rates of change. The rate of change for nonlinear functions is not constant. Therefore, its graphs are not straight lines.

The equation for a linear function can always be written in the form $y = mx + b$, where m represents the constant rate of change. You can determine whether a function is linear by examining its equation. In a linear function, the power of x is always 1 or 0, and x does not appear in the denominator of a fraction.

Example 1 **Determine whether the graph represents a *linear* or *nonlinear* function. Explain.**

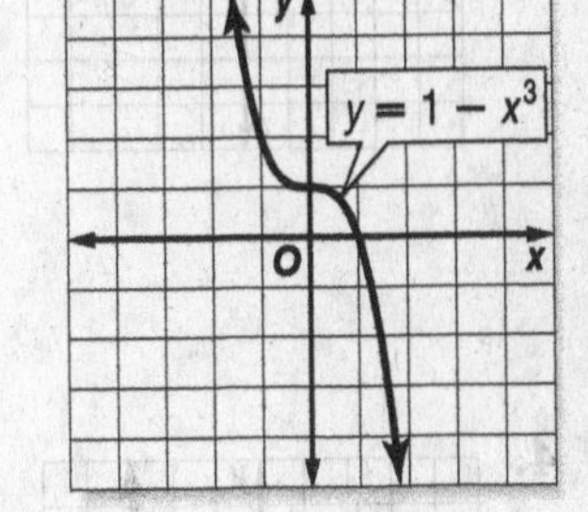

The graph is a curve, not a straight line. So, it represents a nonlinear function.

Example 2 **Determine whether $y = 2.5x$ represents a *linear* or *nonlinear* function. Explain.**

Since the equation can be written as $y = 2.5x + 0$, the function is linear.

A nonlinear function does not increase or decrease at the same rate. You can use a table to determine if the rate of change is constant.

Example 3 **Determine whether the table represents a *linear* or *nonlinear* function. Explain.**

x	−2	2	6	10
y	8	3	−1	−4

(x changes: +4, +4, +4; y changes: −5, −4, −3)

As x increases by 4, y decreases by a different amount each time. The rate of change is not constant, so this function is nonlinear.

Exercises

Determine whether each graph, equation, or table represents a *linear* or *nonlinear* function. Explain.

1. 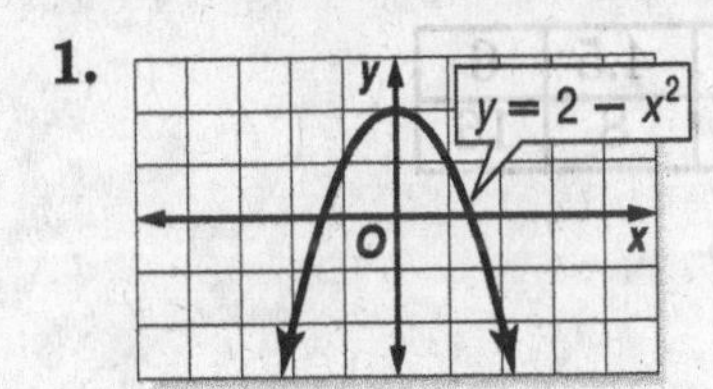

2. 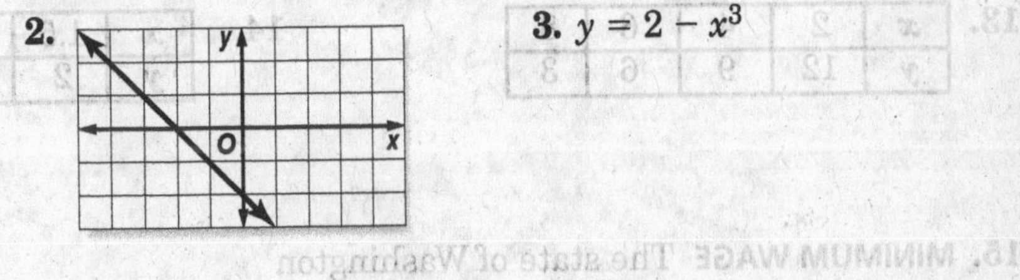

3. $y = 2 - x^3$

4. $y = 5 - 2x$

5.

x	1	2	3	4
y	3	6	9	12

6.

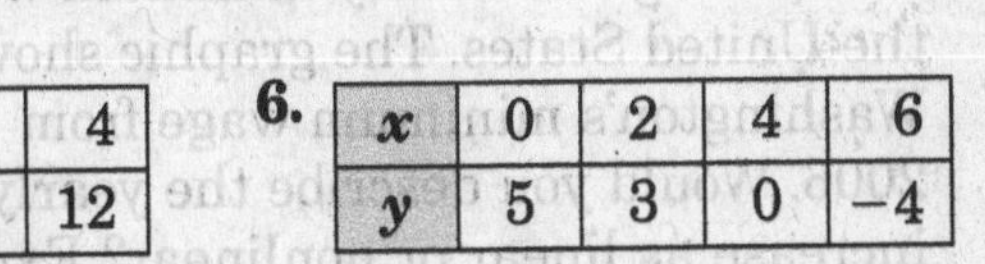

x	0	2	4	6
y	5	3	0	−4

NAME ____ DATE ____________ PERIOD _____

10-1 Practice

Linear and Nonlinear Functions

Determine whether each graph, equation, or table represents a linear or nonlinear function. Explain.

1.

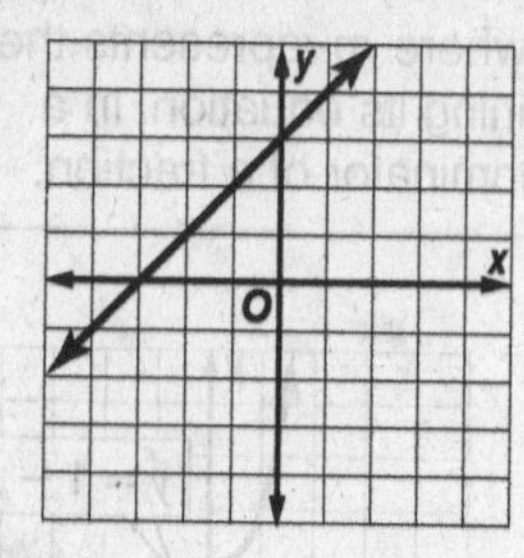

2.

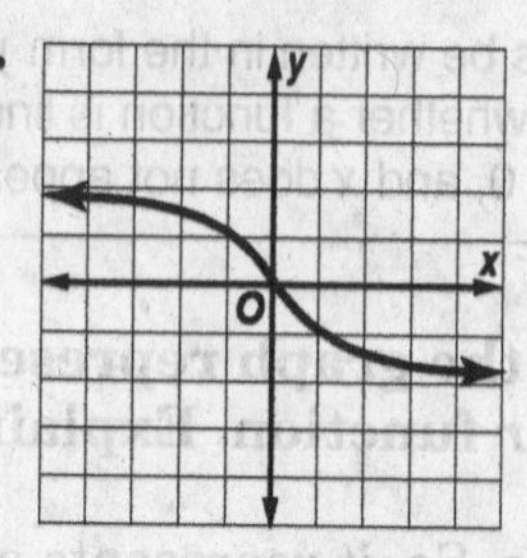

3.

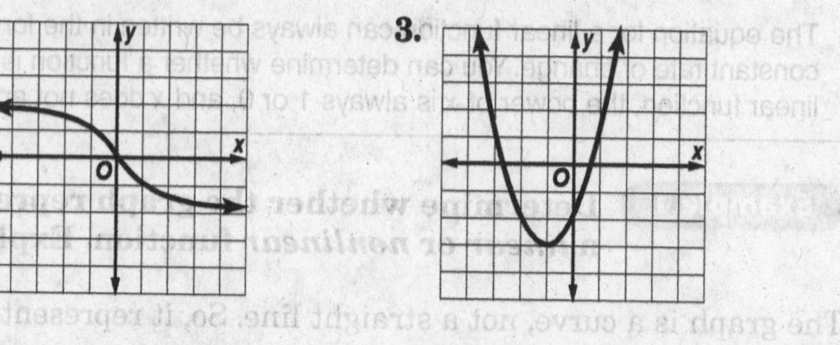

4.

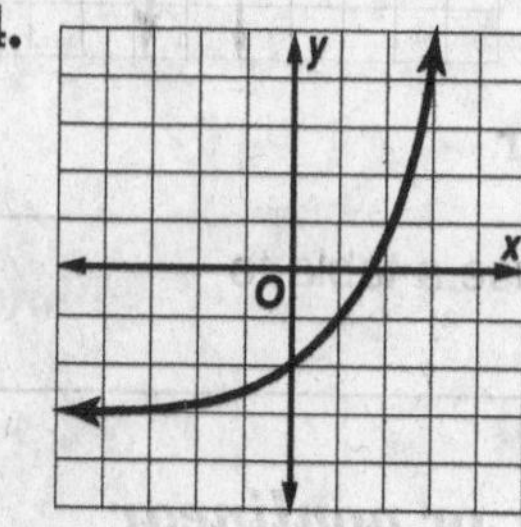

5.

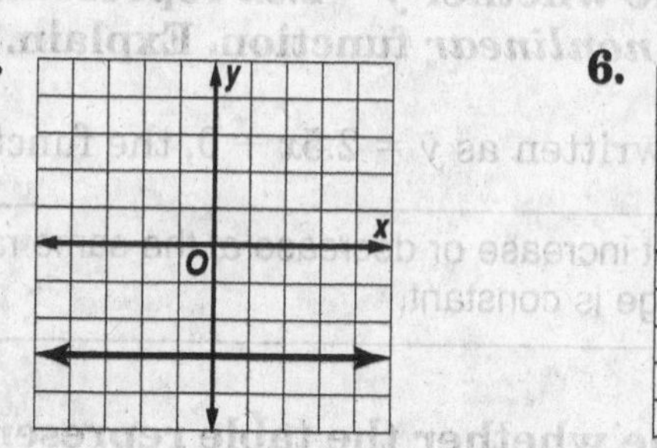

6.

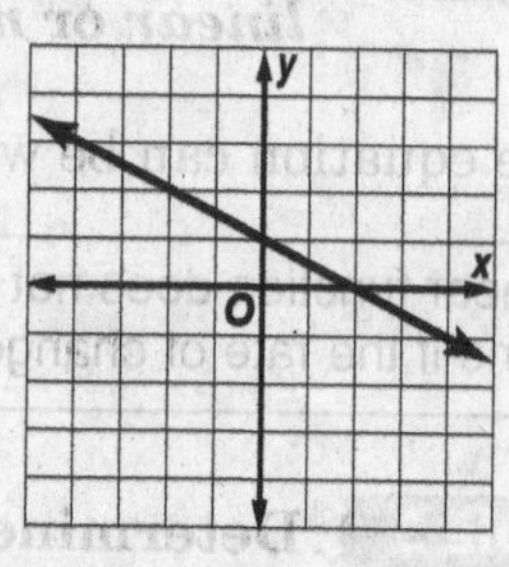

7. $y = 9 - x^2$

8. $y = -2.3x$

9. $y = \frac{x}{9}$

10. $y = \frac{3}{x}$

11. $2x + 3y = 6$

12. $2xy = 12$

13.

x	2	4	6	8
y	12	9	6	3

14.

x	1.5	3	4.5	6
y	2	4	8	16

15. MINIMUM WAGE The state of Washington has the highest hourly minimum wage in the United States. The graphic shows Washington's minimum wage from 1999 to 2006. Would you describe the yearly increase as linear or nonlinear? Explain your reasoning.

10-2 Study Guide and Intervention

Graphing Quadratic Functions

A **quadratic function**, such as $A = s^2$, is a function in which the greatest power of the variable is 2. Its graph is U-shaped, opening upward or downward.

Example 1 Graph $y = -0.25x^2$.

x	$-0.25x^2$	y	(x, y)
-4	$-0.25(-4)^2 = -4$	-4	$(-4, -4)$
-2	$-0.25(-2)^2 = -1$	-1	$(-2, -1)$
0	$-0.25(0)^2 = 0$	0	$(0, 0)$
2	$-0.25(2)^2 = -1$	-1	$(2, -1)$
4	$-0.25(4)^2 = -4$	-4	$(4, -4)$

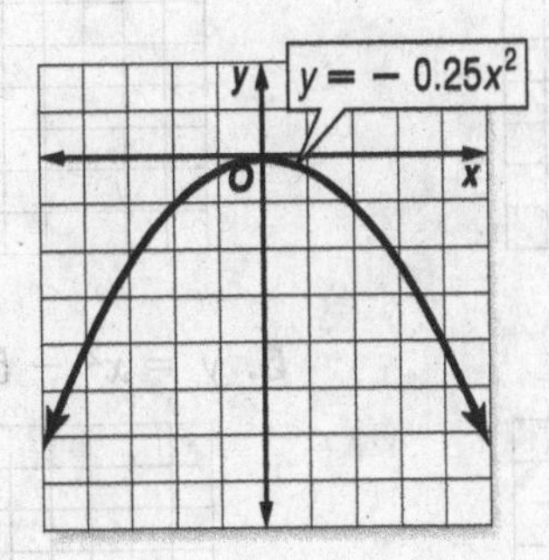

Example 2 Graph $y = x^2 - 3$.

x	$x^2 - 3$	y	(x, y)
-2	$(-2)^2 - 3 = 1$	1	$(-2, 1)$
-1	$(-1)^2 - 3 = -2$	-2	$(-1, -2)$
0	$(0)^2 - 3 = -3$	-3	$(0, -3)$
1	$(1)^2 - 3 = -2$	-2	$(1, -2)$
2	$(2)^2 - 3 = 1$	1	$(2, 1)$

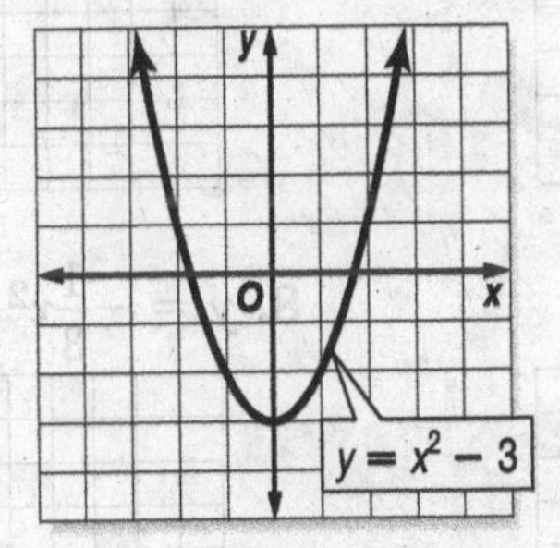

Exercises

Graph each function.

1. $y = 2x^2$

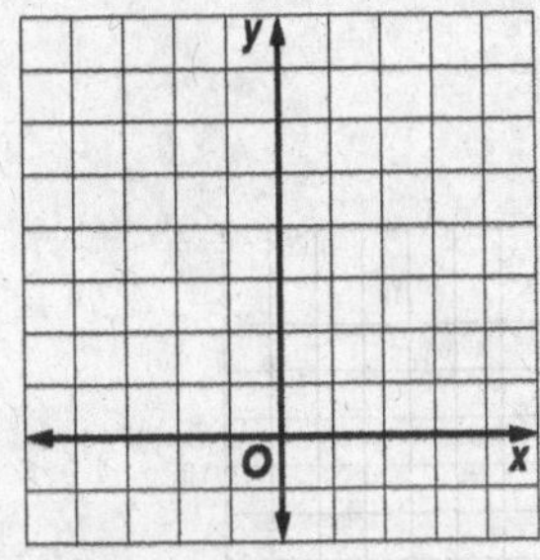

2. $y = -0.5x^2$

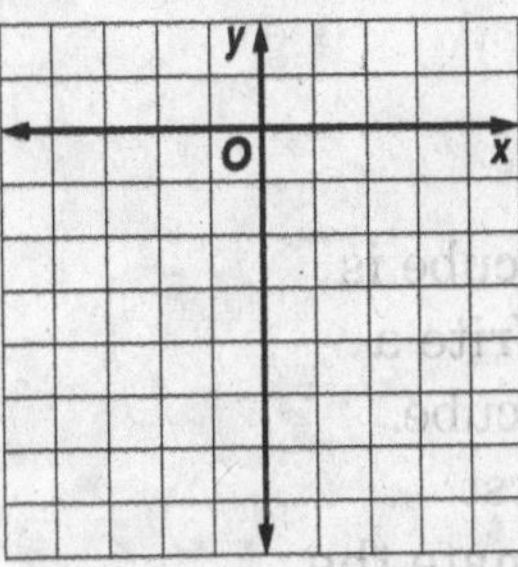

3. $y = x^2 - 1$

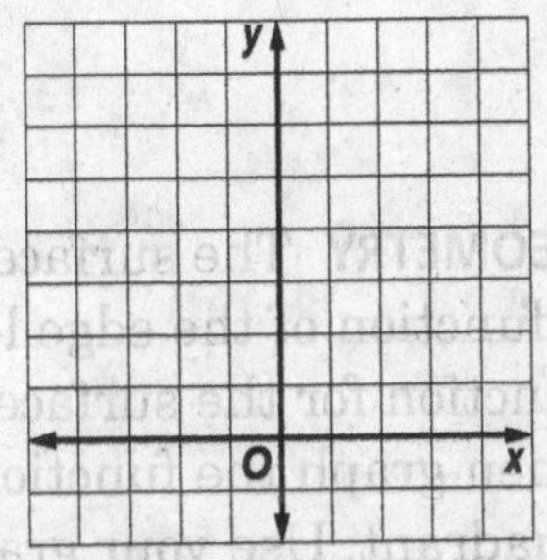

4. $y = 2x^2 + 4$

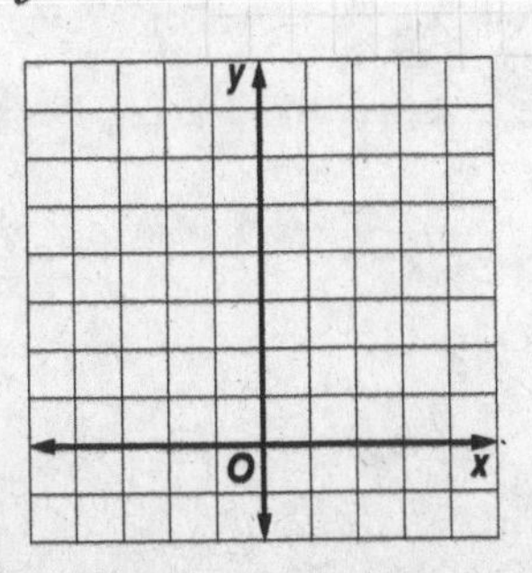

5. $y = -x^2 - 3$

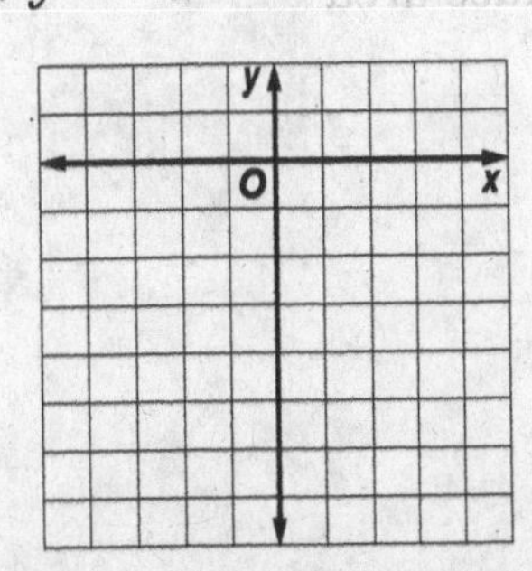

6. $y = -3x^2 + 1$

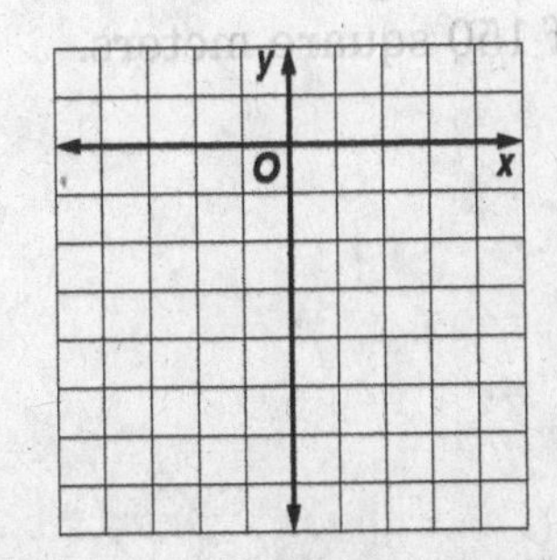

NAME ______________________________ DATE ____________ PERIOD _____

10-2 Practice

Graphing Quadratic Functions

Graph each function.

1. $y = x^2$

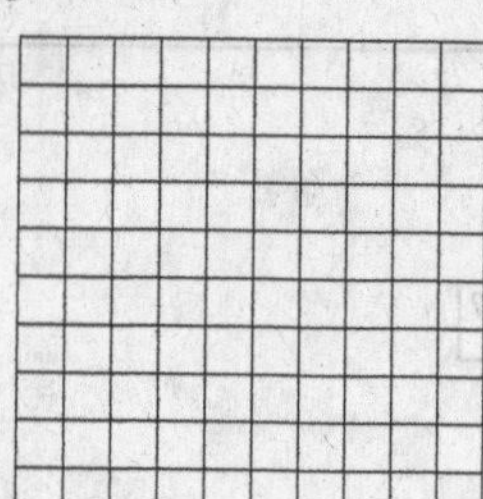

2. $y = -x^2$

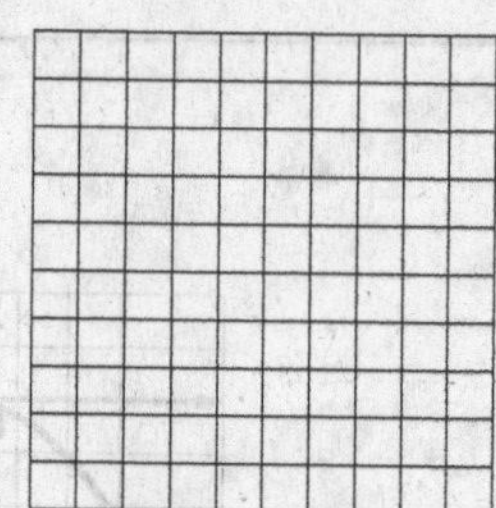

3. $y = x^2 + 3$

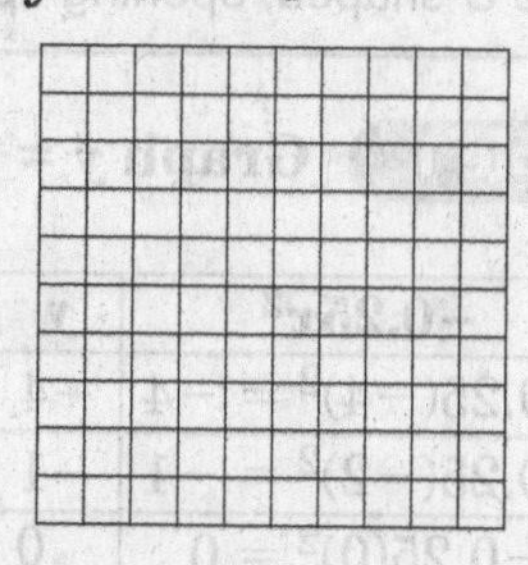

4. $y = -x^2 + 3$

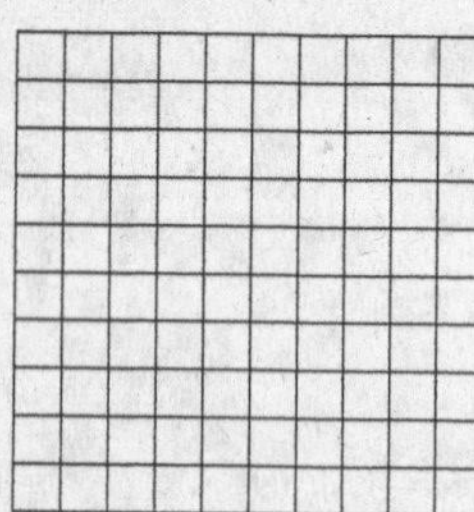

5. $y = x^2 - 5$

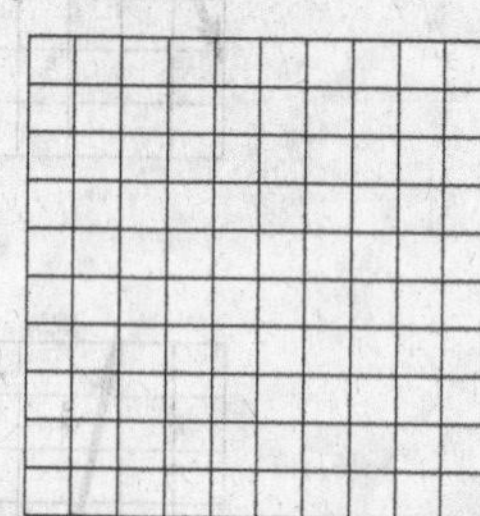

6. $y = 3x^2 - 4$

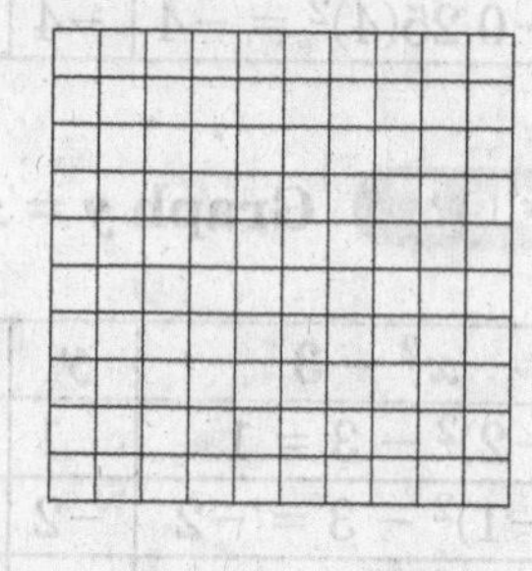

7. $y = 2.5x^2 - 3$

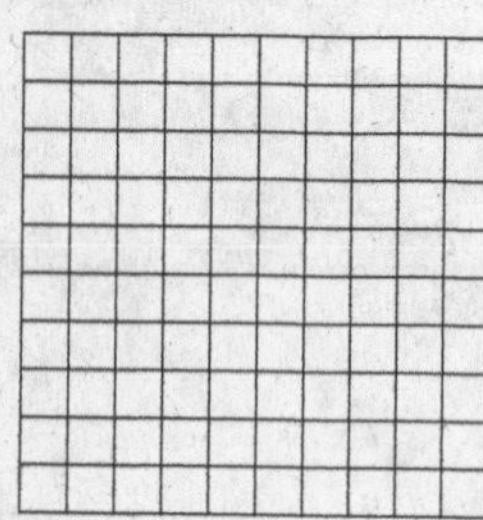

8. $y = -\frac{1}{3}x^2$

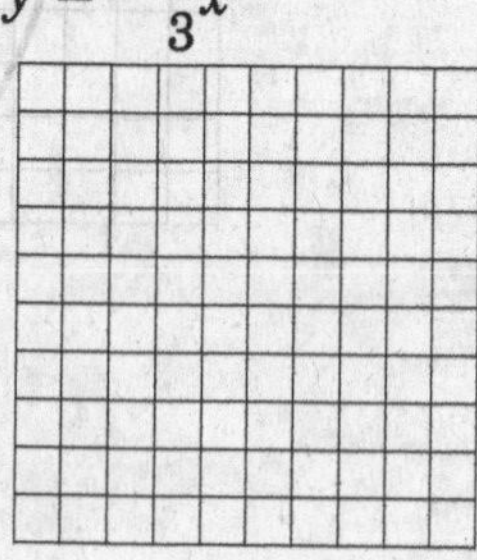

9. $y = -\frac{1}{2}x^2 + 2$

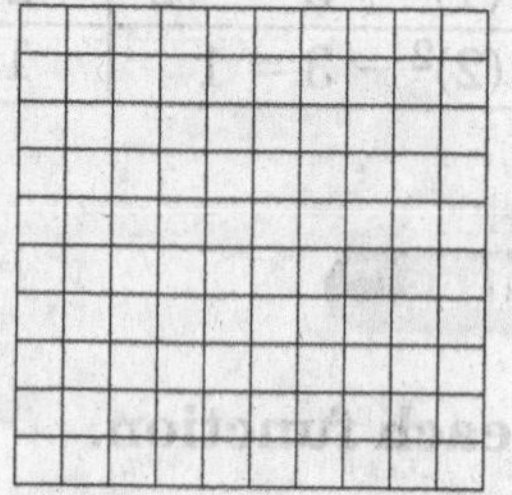

10. GEOMETRY The surface area of a cube is a function of the edge length a. Write a function for the surface area of a cube. Then graph the function in the first quadrant. Use your graph to estimate the edge length of a cube with a surface area of 150 square meters.

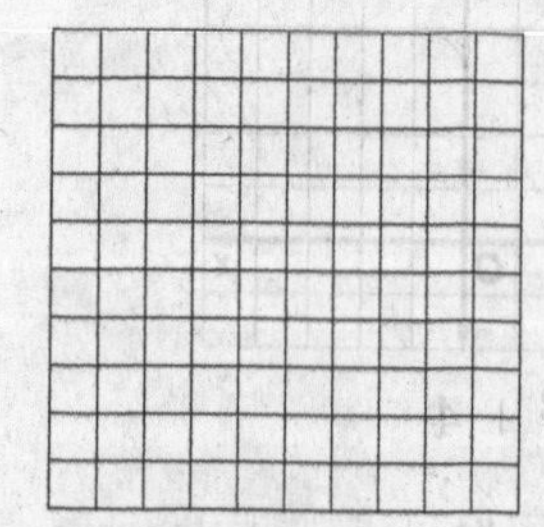

10-3 Study Guide and Intervention

Problem-Solving Investigation: Make a Model

You may need to use the make-a-model strategy to solve some problems.

You can always use the four-step plan to solve a problem.

Understand	• Determine what information is given in the problem and what you need to find.
Plan	• Select a strategy including a possible estimate.
Solve	• Solve the problem by carrying out your plan.
Check	• Examine your answer to see if it seems reasonable.

Example

Kisha is trying to make a box out of a piece of cardboard by cutting a square out of each corner. She will then fold up the sides and tape them together. The cardboard measures 4 feet 6 inches by 6 feet 6 inches. She wants the box to measure 3 feet wide by 5 feet long. What size squares should Kisha cut out of the corners to make the box?

Understand She wants to know what size squares to cut out of each corner to make a box which measures 3 feet by 5 feet by 9 inches.

Plan Start by making a model of the cardboard. Label the sides of the cardboard in feet. Draw lines to show the squares that will be cut out of the corners.

Solve

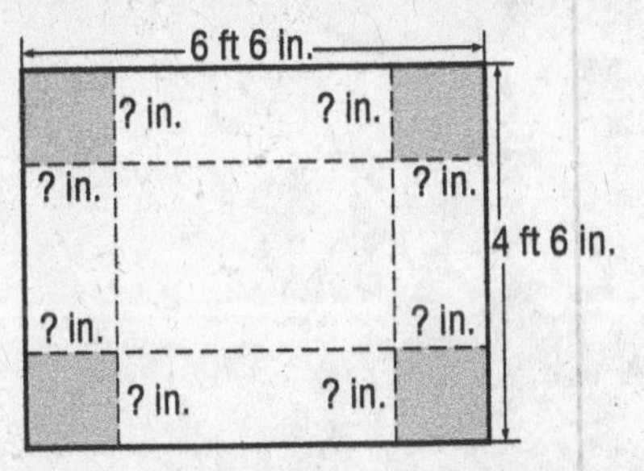

Subtract 5 feet from 6 feet 6 inches and divide by 2.
1 ft 6 in. = 18 in. 18 in. ÷ 2 = 9 in.

The square must have sides that are 9 inches long.

Check Check that the width of the box meets the specifications. Subtracting 18 inches or 1 foot 6 inches from 4 feet 6 inches yields 3 feet, which is the width required.

Exercises

Make a model to solve each problem.

1. **CONSTRUCTION** A chicken coop will be 20 feet long and 16 feet wide. One side that is 20 feet long will be formed by the barn. The other three sides will be made of wire fencing with posts at every corner and every 4 feet between each corner. How many feet of fencing and how many posts are needed to build the chicken coop?

2. **GEOMETRY** What is the fewest number of one-inch cubes needed to make a rectangular prism that measures 4 inches by 5 inches by 6 inches? (Hint: The prism can be hollow inside.)

NAME ______________________________ DATE ____________ PERIOD ______

10-3 Practice

Problem-Solving Investigation: Make a Model

Mixed Problem Solving

For Exercises 1 and 2, solve using the make-a-model strategy.

1. **QUILTS** Mrs. Renoir has completed the interior portion of a quilt top measuring 4 feet by 6 feet. She is outlining this with squares measuring 4 inches on each side. How many such squares will she need?

2. **GEOMETRY** Kumiko has four plastic shapes, a circle, a square, a triangle, and a pentagon. In how many ways can she line up the four shapes if the circle cannot be next to the square?

Use any strategy to solve Exercises 3–6. Some strategies are shown below.

PROBLEM-SOLVING STRATEGIES
• Draw a diagram.   • Guess and check.   • Make a model.

3. **FOOTBALL** The attendance at the first two football games of the season are shown in the table. Did the attendance increase by about 1% or about 10%?

Football Attendance

Game 1	5,049
Game 2	5,582

4. **GAMES** Jamal has a deck of 40 cards. After giving each player in the game an equal number of cards, he has four cards left over, which is not enough to give each player another card. How many players could be in the game?

5. **CLOTHING** Salome has 5 T-shirts, 3 pairs of jeans, and 2 pairs of sneakers. In how many ways can she choose one T-shirt, one pair of jeans, and one pair of sneakers to wear today?

6. **NUMBER THEORY** After adding 8 to a number and then dividing by 3, the result is 19. What is the number?

NAME ______________________ DATE ____________ PERIOD ______

10-4 Study Guide and Intervention

Graphing Cubic Functions

A **cubic function,** such as $A = s^3$, is a function in which the greatest power of the variable is 3. Its graph is a curve. You can graph cubic functions by making a table of values.

Example 1 Graph $y = -x^3$.

x	$y = -x^3$	(x, y)
-2	$-(-2)^3$	$(-2, 8)$
-1	$-(-1)^3$	$(-1, -1)$
0	$-(0)^3$	$(0, 0)$
1	$-(1)^3$	$(1, -1)$
2	$-(2)^3$	$(2, -8)$

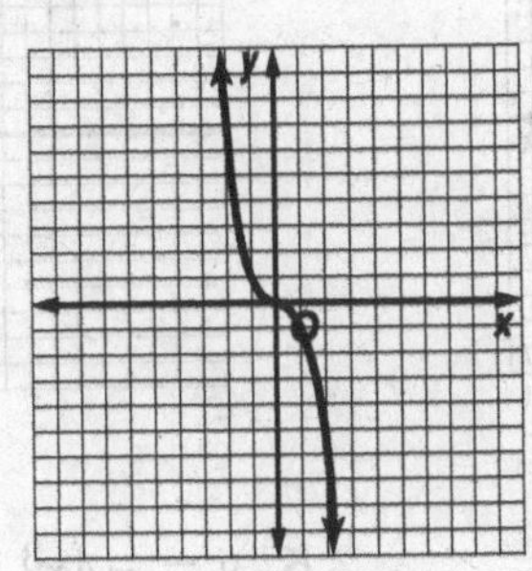

Example 2 Graph $y = x^3$.

x	$y = x^3$	(x, y)
-1.5	$(-1.5)^3 \approx -3.4$	$(-1.5, -3.4)$
-1	$(-1)^3 = -1$	$(-1, -1)$
0	$-(0)^3 = 0$	$(0, 0)$
1	$(1)^3 = 1$	$(1, 1)$
1.5	$(1.5)^3 \approx 3.4$	$(1.5, 3.4)$

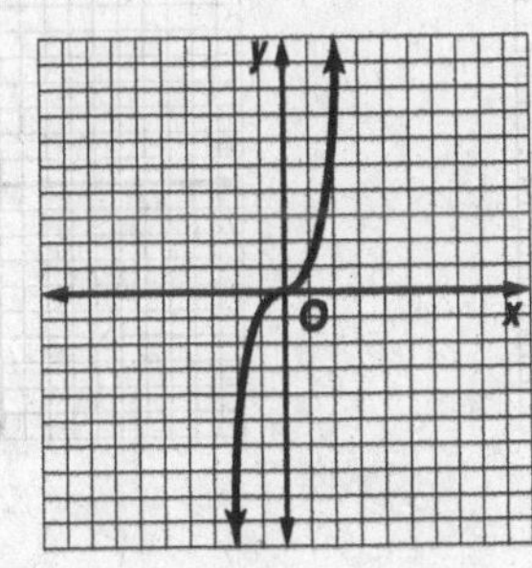

Exercises

Graph each function.

1. $y = x^3 + 1$

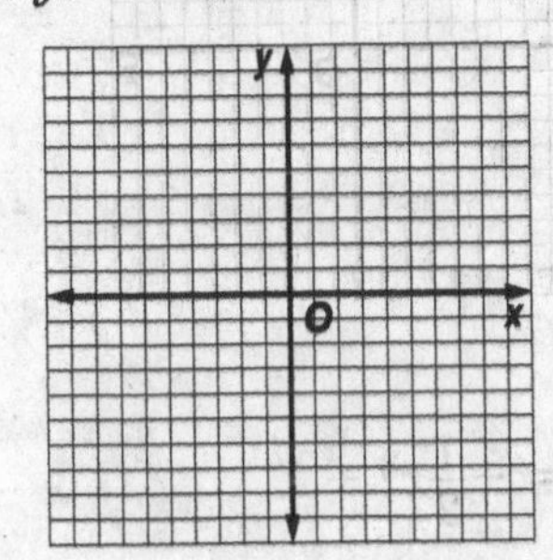

2. $y = x^3 - 2$

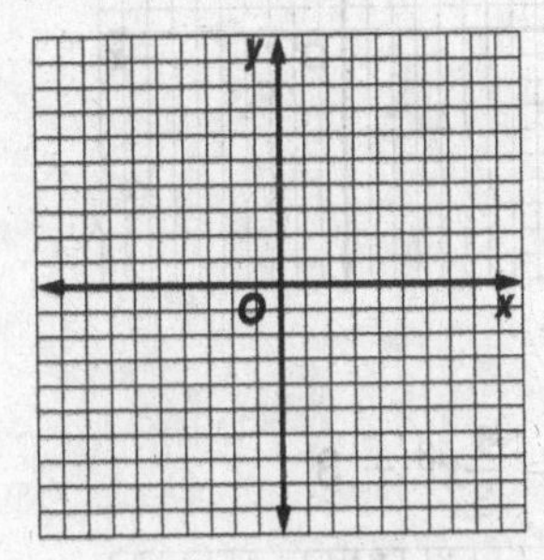

3. $y = 2x^3$

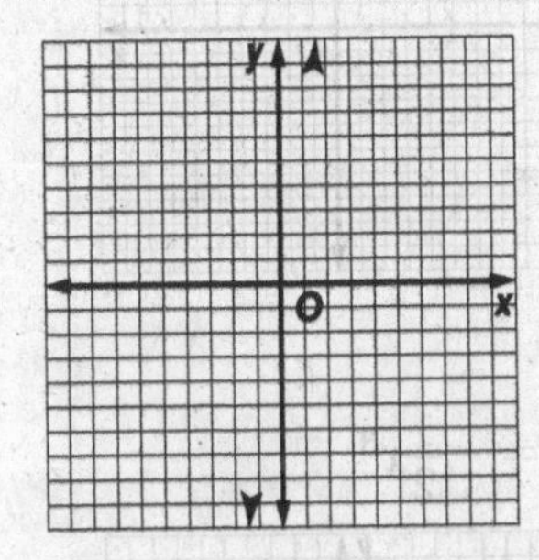

4. $y = 2x^3 - 1$

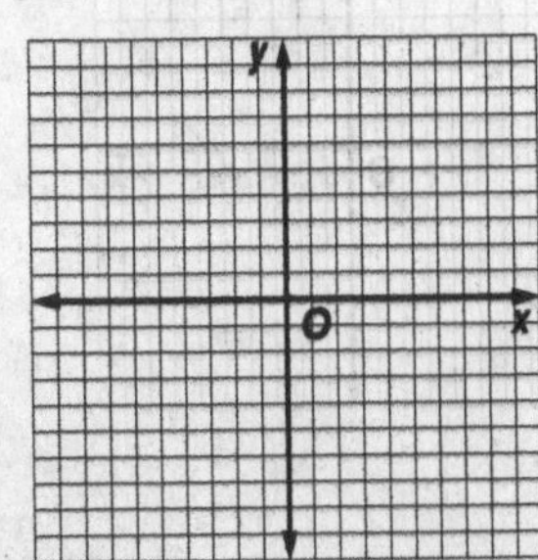

5. $y = 3x^3 + 1$

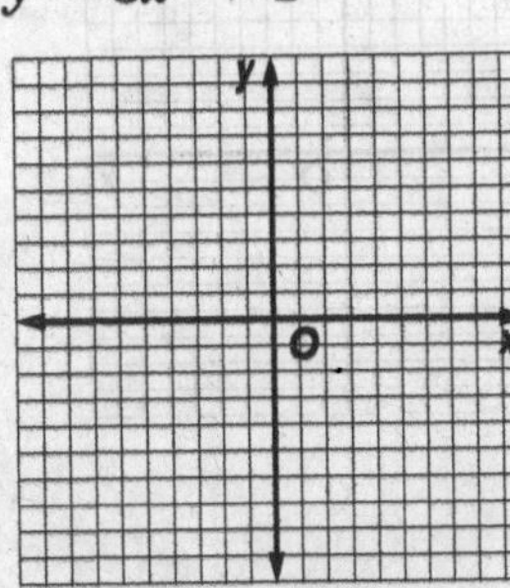

6. $y = -x^3 - 1$

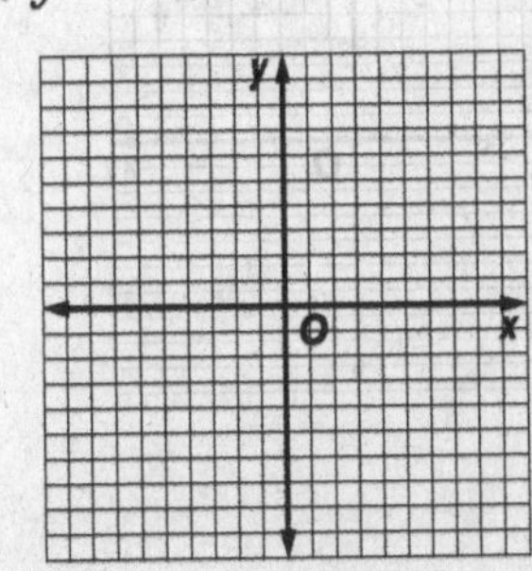

Lesson 10-4

NAME ______________________ DATE __________ PERIOD _____

10-4 Practice

Graphing Cubic Functions

Graph each function.

1. $y = 0.1x^3 + 1$

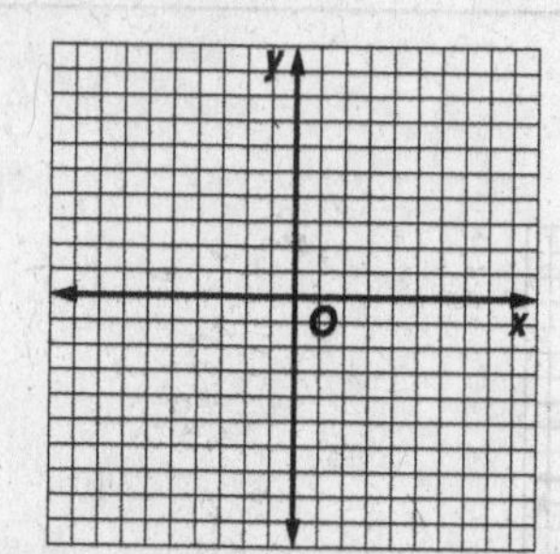

2. $y = \frac{1}{4}x^3$

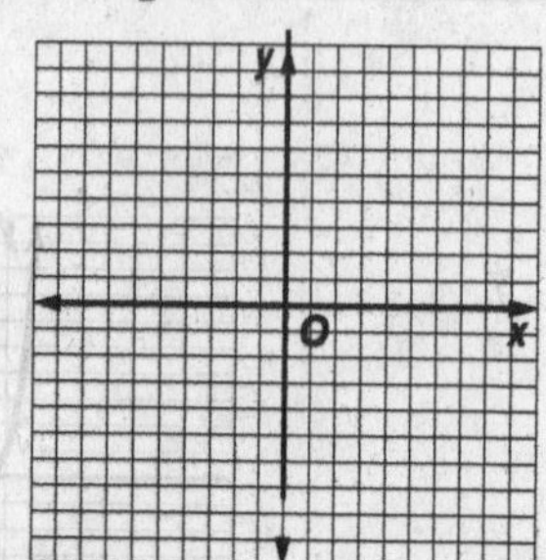

3. $y = 0.2x^3$

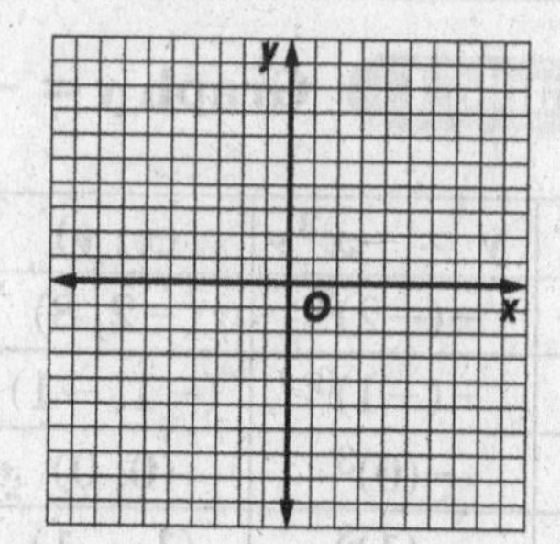

4. $y = \frac{1}{3}x^3$

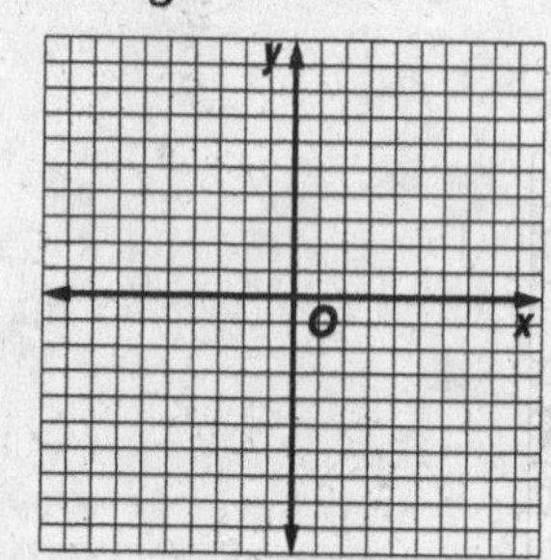

5. $y = -4x^3$

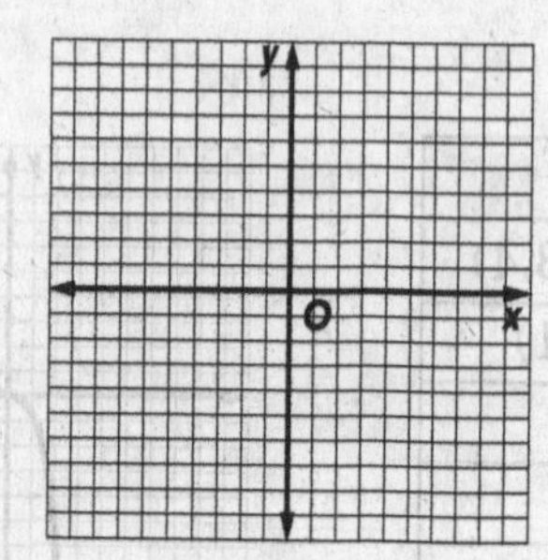

6. $y = \frac{1}{2}x^3$

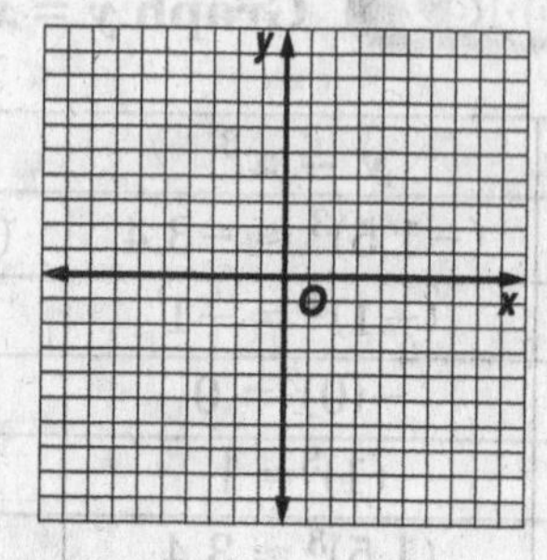

7. $y = -\frac{1}{4}x^3$

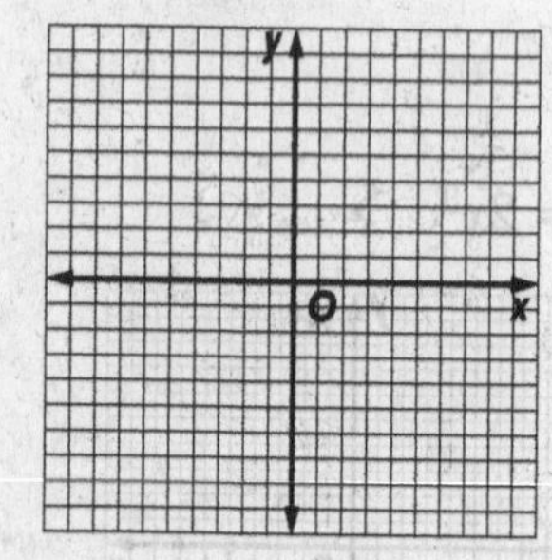

8. $y = -0.2x^3$

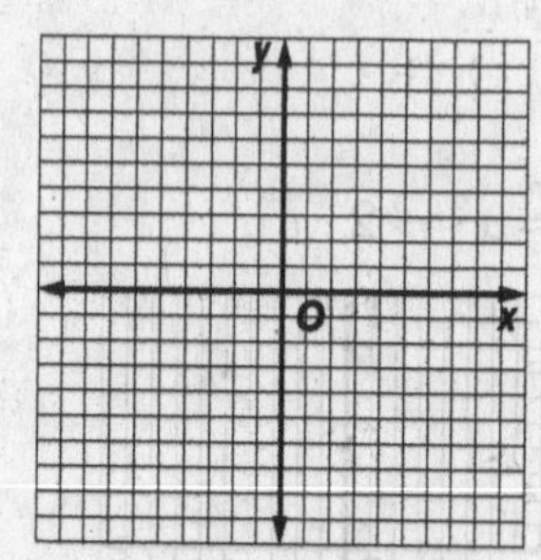

9. $y = 0.1x^3$

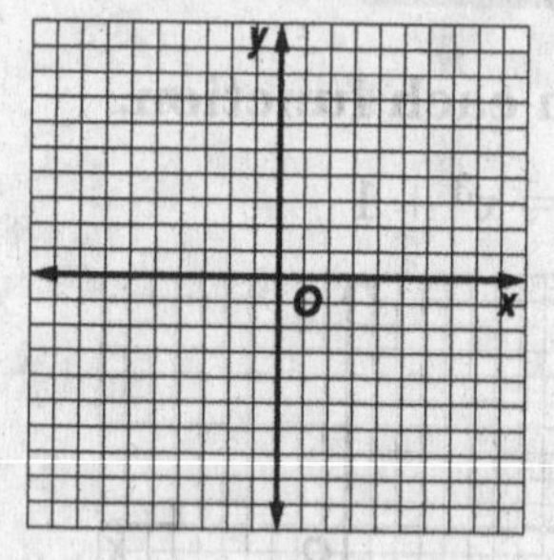

10. $y = -\frac{1}{2}x^3$

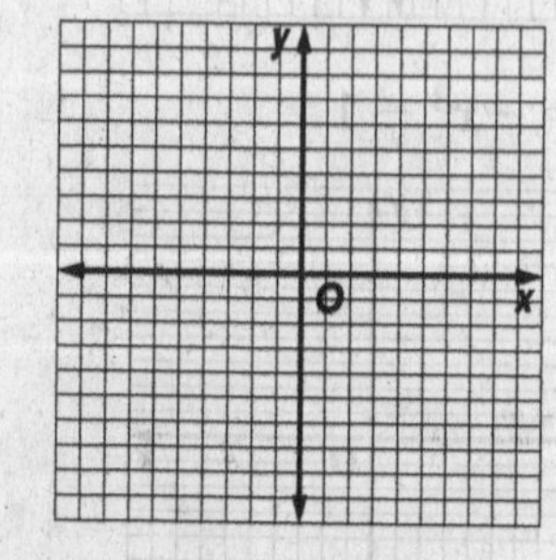

11. $y = \frac{1}{3}x^3 + 3$

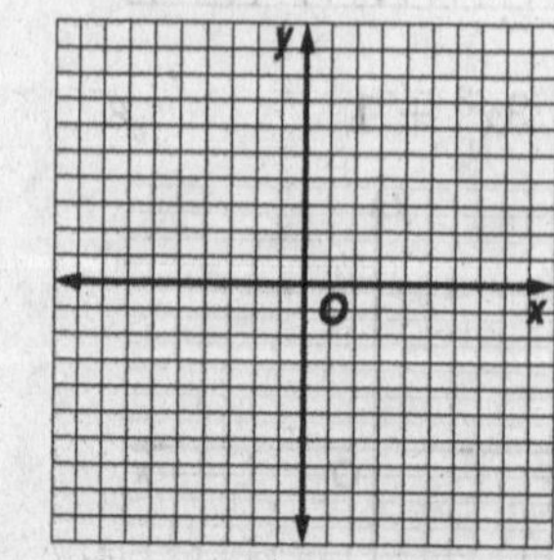

12. $y = -\frac{1}{2}x^3 + 1$

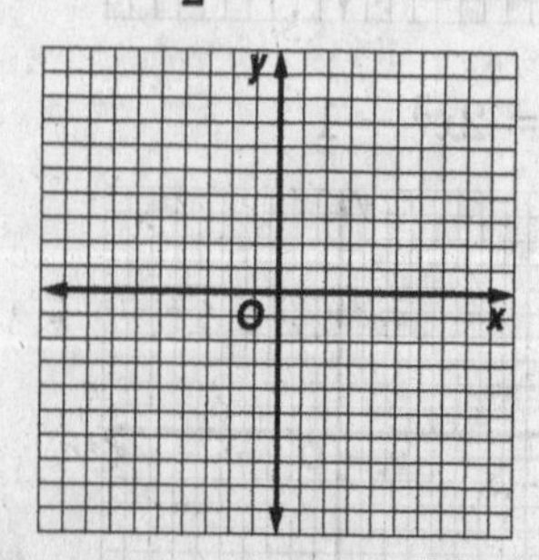

NAME ______________________________ DATE ____________ PERIOD ______

10-5 Study Guide and Intervention

Multiplying Monomials

The **Product of Powers Property** states that to multiply powers that have the same base, add the exponents: $a^n \cdot a^m = a^{n+m}$.

Example **Multiply. Express using exponents.**

1 $\mathbf{2^3 \cdot 2^2}$

$2^3 \cdot 2^2 = 2^{3+2}$ — The common base is 2.

$= 2^5$ — Add the exponents.

2 $\mathbf{-2s^6(-7s^7)}$

$-2s^6(-7s^7) = (-2 \cdot -7)(s^6 \cdot s^7)$ — Commutative and Associative Properties

$= (14)(s^{6+7})$ — The common base is s.

$= 14s^{13}$ — Add the exponents.

3 $\mathbf{n^5 \cdot n^{-3}}$

$n^5 \cdot n^{-3} = n^{5-3}$ — The common base is n.

$= n^2$ — Subtract the exponents.

Exercise

Multiply. Express using exponents.

1. $3^4 \cdot 3^1$

2. $5^2 \cdot 5^5$

3. $e^2 \cdot e^7$

4. $2a^5 \cdot 6a$

5. $-3t^3 \cdot 2t^8$

6. $4x^2(-5x^6)$

7. $-6t^4 \cdot -3t^5$

8. $\left(\frac{3}{4}\right)^{-3} \cdot \left(\frac{3}{4}\right)^{6}$

9. $-6m^2 \cdot 4m$

10. $3s^6(-9s^{-2}h^2)$

11. $9a^2(-6a^{-5})$

12. $-2e^4z^{-4}(6e^{-6})$

NAME ______________________ DATE __________ PERIOD _____

10-5 Practice

Multiplying Monomials

Multiply. Express using exponents.

1. $5^9 \cdot 5^3$

2. $3^8 \cdot 3$

3. $c \cdot c^6$

4. $m^5 \cdot m^2$

5. $3x \cdot 4x^4$

6. $(2h^7)(7h)$

7. $-5d^6(8d^6)$

8. $(6k^5)(-k^4)$

9. $(-w)(-10w^3)$

10. $-7z^4(-3z^8)$

11. $bc^3(b^2c)$

12. $3a^4 \cdot 6a^2$

13. $3m^3n^2(8mn^3)$

14. $\frac{3}{7} \times \left(\frac{3}{7}\right)^{-1}$

15. $(-4k^{-5})(5k^{-6})$

16. $7t^5(-6t^5)$

17. $(-6d^{-4})(-9d^5e^{-5})$

18. $(3s^5)(-4s^{-6})$

19. $(-8b^{-2})(2b^{-3})$

20. $\left(\frac{2}{9}\right)^{-2} \times \left(\frac{2}{9}\right)^{-4}$

21. $(3ab^2)(a^2c^5)$

22. $(9p^4)(-8p^2)$

23. $(6f^5)(-f^{-2}g^3)$

24. $-10v^4(2v^5)$

25. **CAR LOANS** After making a down payment, Mr. Valle will make 6^2 monthly payments of 6^3 dollars each to pay for his new car. What is the total of the monthly payments?

26. **MEASUREMENT** Find the area of a rectangle with a width of $4x$ and a length of $6x$.

10-6 Study Guide and Intervention

Dividing Monomials

The Quotient of Powers Property states that to divide powers that have the same base, subtract the exponents: $a^n \div a^m = a^{n-m}$.

Example **Divide. Express using exponents.**

1 $\dfrac{k^8}{k}$

$\dfrac{k^8}{k} = k^{8-1}$ — The common base is *k*.

$= k^7$ — Subtract the exponents.

2 $\dfrac{28g^{12}}{-4g^3}$

$\dfrac{28g^{12}}{-4g^3} = \left(\dfrac{28}{-4}\right)\left(\dfrac{g^{12}}{g^3}\right)$ — Commutative and Associative Properties

$= (-7)(g^{12-3})$ — The common base is *g*.

$= -7g^9$ — Subtract the exponents.

3 $\dfrac{5^8}{5^{-5}}$

$\dfrac{5^8}{5^{-5}} = 5^{8-(-5)}$ — Quotient of Powers.

$= 5^{13}$ — Simplify.

Exercise

Divide. Express using exponents.

1. $\dfrac{2^8}{2^6}$ **2.** $\dfrac{7^9}{7^3}$ **3.** $\dfrac{v^{14}}{v^6}$

4. $\dfrac{15w^7}{5w^2}$ **5.** $\dfrac{21z^{10}}{7z^9}$ **6.** $\dfrac{10m^8}{2m}$

7. $\dfrac{(-12)^3}{(-12)^3}$ **8.** $\dfrac{c^{20}}{c^{13}}$ **9.** $\dfrac{1^8}{1^6}$

10. $\dfrac{x^{-2}}{x^{-4}}$ **11.** $\dfrac{100^7}{100^6}$ **12.** $\dfrac{4^{-2}}{4^6}$

NAME ______________________ DATE ____________ PERIOD _____

10-6 Practice

Dividing Monomials

Divide. Express using exponents.

1. $\frac{5^9}{5^3}$
2. $\frac{3^8}{3}$
3. $\frac{b^6}{b^4}$
4. $\frac{g^{15}}{g^7}$
5. $\frac{18v^5}{9v}$
6. $\frac{24a^6}{6a^5}$
7. $\frac{30s^7t}{5st}$
8. $\frac{28p^5q^4}{14p^4q^3}$
9. $(9j^7k^4) \div (3j^6k)$
10. $(32x^9y^5) \div (16x^5y^3)$
11. $\frac{\left(\frac{2}{7}\right)^2 \times \left(\frac{2}{7}\right)^{-5}}{\left(\frac{2}{7}\right)^3}$
12. $\frac{4b^5}{2b^{-8}}$
13. $\frac{5^8 \cdot 5^2 \cdot 5^4}{5^7 \cdot 5^4 \cdot 5^6}$
14. $\frac{36^{-6}}{36^2} \times 36^{-8}$
15. $\frac{p}{p^4}$
16. $\frac{7^{-3}}{7^{-6}}$
17. $\frac{\left(\frac{1}{3}\right)^{-2} \times \left(\frac{1}{3}\right)}{\left(\frac{1}{3}\right)^2}$
18. $\frac{95^{21}}{95^{18}}$
19. $\frac{v^{20}}{v^{30}}$
20. $(8b^6d^9) \div (2b^2d^3)$
21. $\frac{n^{19}}{n^{11}}$
22. $\frac{32s^6t^4}{16s^4t^3}$
23. $y^6 \div y^3$
24. $\frac{a^2b^2}{b^2}$

25. **BONUSES** A company has set aside 10^7 dollars for annual employee bonuses. If the company has 10^4 employees and the money is divided equally among them, how much will each employee receive?

26. **ACIDITY** For each increase of one in pH level, the acidity of a substance is 10 times greater. The pH level of baking soda is 8, and the pH level of lye is 13. How many times greater is the alkalinity of lye than that of baking soda?

NAME ______________________ DATE __________ PERIOD _____

10-7 Study Guide and Intervention

Powers of Monomials

Rule: To find the power of a power, multiply the exponents.

Rule: To find the power of a product, find the power of each factor and multiply.

Example 1 **Simplify to find the power of the power:**

$(5^3)^6$

$(5^3)^6 = 5^{3 \cdot 6}$

$= 5^{18}$

Example 2 **Simplify to find the power of each factor.**

$(-3m^2n^4)^3$

$(-3m^2n^4)^3 = (-3)^3 \cdot m^{2 \cdot 3} \cdot n^{4 \cdot 3}$

$= -27m^6n^{12}$

Exercise

Simplify to find the power of the power.

1. $(4^3)^5$

2. $(4^2)^7$

3. $(9^2)^4$

4. $(k^4)^2$

5. $[(6^3)^2]^2$

6. $[(3^2)^2]^3$

Simplify to find the power of each product.

7. $(5q^4r^2)^5$

8. $(3y^2z^2)^6$

9. $(7a^4b^3c^7)^2$

10. $(-4d^3e^5)^2$

11. $(-5g^4h^9)^7$

12. $(0.2k^8)^2$

Lesson 10-7

10-7 Practice

Powers of Monomials

Simplify.

1. $(6t^5)^2$ **2.** $(4w^9)^4$ **3.** $(12k^6)^3$ **4.** $(15m^8)^3$

5. $(4d^3e^5)^7$ **6.** $(-4r^6s^{15})^4$ **7.** $[(7^2)^2]^2$ **8.** $[(3^2)^2]^3$

9. $(\frac{3}{5}a^{-6}b^9)^2$ **10.** $(4x^{-2})^3(3x^6)^4$ **11.** $(0.6p^5)^3$ **12.** $\left(\frac{1}{5}w^5z^3\right)^2$

GEOMETRY **Express the area of each square below as a monomial.**

13.

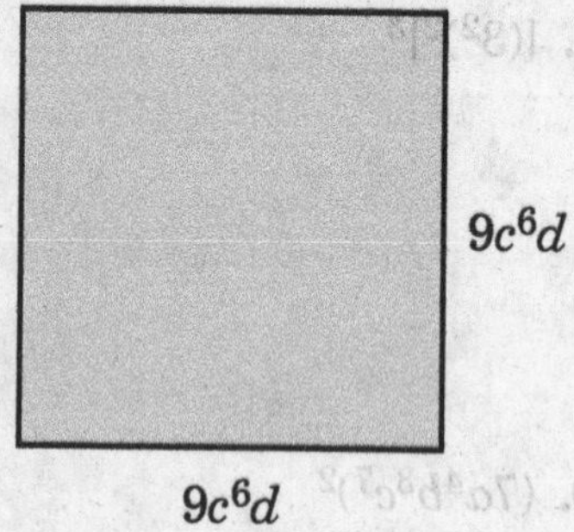

14.

$14g^5h^9$

$14g^5h^9$

15. MEASUREMENT In the Metric System, you would need to have $(10^4)^2$ grams just to equal 1 metric ton. Simplify this measurement by multiplying the exponents, then simplify by finding the actual number of grams needed to equal 1 metric ton.

NAME ______________________ DATE __________ PERIOD _____

10-8 Study Guide and Intervention

Roots of Monomials

The **square root** of a monomial is one of the two equal factors of the monomial.

Example 1 **Simplify $\sqrt{25a^4}$.**

$\sqrt{25a^4} = \sqrt{25} \cdot \sqrt{a^4}$ Product Property of Square Roots

$= 5 \cdot a^2$ ← Absolute value is not necessary since the value of a^2 will never be negative.

Example 2 **Simplify $\sqrt{49y^6z^8}$.**

$\sqrt{49y^6z^8} = \sqrt{49} \cdot \sqrt{y^6} \cdot \sqrt{z^8}$ Product Property of Square Roots

$= 7 \cdot |y^3| \cdot z^4$ ← Use absolute value to indicate the positive value of y^3.

The **cube root** of a monomial is one of the three equal factors of the monomial.

Example 3 **Simplify $\sqrt[3]{d^6}$.**

$\sqrt[3]{d^6} = d^2$ $(d^2)^3 = d^6$

Example 4 **Simplify $\sqrt[3]{125m^9n^{12}}$.**

$\sqrt[3]{125m^9n^{12}} = \sqrt[3]{125} \cdot \sqrt[3]{m^9} \cdot \sqrt[3]{n^{12}}$ Product Property of Cube Roots

$= 5 \cdot m^3 \cdot n^4$ $(5)^3 = 125$; $(m^3)^3 = m^9$; and $(n^4)^3 = n^{12}$

Exercises

Simplify.

1. $\sqrt{c^2}$ **2.** $\sqrt{4s^6}$ **3.** $\sqrt{16a^8b^{12}}$

4. $\sqrt{64g^8h^{10}}$ **5.** $\sqrt{36r^2s^6}$ **6.** $\sqrt{121d^4e^{10}}$

7. $\sqrt[3]{p^6}$ **8.** $\sqrt[3]{27m^{15}}$ **9.** $\sqrt[3]{216a^9b^{21}}$

10. $\sqrt[3]{64y^{12}z^{24}}$ **11.** $\sqrt[3]{343t^{18}u^6}$ **12.** $\sqrt[3]{125p^{15}q^{27}}$

Lesson 10-8

NAME ______________________ DATE ____________ PERIOD ______

10-8 Practice

Roots of Monomials

Simplify.

1. $\sqrt{n^6}$

2. $\sqrt{y^8}$

3. $\sqrt{169p^4}$

4. $\sqrt{0.36a^6b^2}$

5. $\sqrt{2.25f^{10}g^{12}}$

6. $\sqrt{\frac{1}{49}j^4k^6m^8}$

7. $\sqrt[3]{x^9}$

8. $\sqrt[3]{c^{18}}$

9. $\sqrt[3]{27m^{15}}$

10. $\sqrt[3]{0.027t^{12}u^6}$

11. $\sqrt[3]{\frac{1}{64}y^{21}z^{27}}$

12. $\sqrt[3]{0.216e^{12}f^{18}}$

13. **MEASUREMENT** Express the side of one side of a square whose area is $64s^6t^8$ square units as a monomial.

14. **MEASUREMENT** Express the side of one side of a square whose volume is $64x^{15}y^{18}$ cubic units as a monomial.

Write a radical expression for each square root.

15. $6|j^5|k^4$

16. $13|n^7p^{11}|$

17. $10s^8t^{12}$

Write a radical expression for each cube root.

18. $7y^4z^5$

19. $5a^2b^7$

20. $6c^3d^8e^4$

11-1 Study Guide and Intervention

Problem-Solving Investigation: Make a Table

You may need to use the make a table strategy to solve some problems.

Understand Determine what information is given in the problem and what you need to find out.

Plan Select a strategy including a possible estimate.

Solve Solve the problem by carrying out your plan.

Check Examine your answer to see if it seems reasonable.

Example

For his science fair project, August decided to classify the 20 rocks and minerals in his collection by their hardness using the Mohs scale. After performing various tests for hardness, he recorded the hardness value of each rock or mineral in his collection in a list. Organize the data in a table using hardness intervals 1–2, 3–4, 5–6, 7–8, 9–10. What is the most common interval of rock hardness?

2	1	5	3	3	10	2	9	4	7	6	3	4	2	3	3	1	5	6	3

Understand You have a list of the hardness values for each rock or mineral. You need to know how many rocks have a hardness between 1 and 2, 3 and 4, 5 and 6, 7 and 8, and 9 and 10. Then you need to determine the most common interval of hardness.

Plan Make a frequency table with intervals to organize the information.

Hardness Interval	Tally	Frequency
1–2	卌	5
3–4	卌 III	8
5–6	IIII	4
7–8	I	1
9–10	II	2

Solve The most common interval of rock hardness is 3–4.

Check August tested 20 rocks for hardness. Since there are 20 values listed, the table seems reasonable.

Exercises

Make a table to solve each problem.

1. **BANKING** The list shows the amount of cash requested by each person that used a certain Automated Teller Machine (ATM) in one day. What is the most common amount of money requested by ATM users?

   $20 $40 $20 $100 $300 $80 $40 $40 $80 $100 $120 $20
   $40 $80 $100 $60 $60 $20 $80 $100 $40 $20 $80 $40

2. **COFFEE** The list shows the coffee sizes in ounces purchased in one hour at a local coffee house. What is the most commonly purchased size of coffee?

   8 16 16 20 8 12 16 8 12 20
   20 16 12 8 8 16 16 20 16 20

NAME ______________________ DATE __________ PERIOD _____

11-1 Practice

Problem-Solving Investigation: Make a Table

Mixed Problem Solving

Use the make a table strategy to solve Exercises 1 and 2.

1. **LIZARDS** Biologists recorded the lengths of lizards they found in the desert. About what percent of the lizard lengths are from 3.0 to 6.9 inches?

Lengths of Lizards Found		
Length (in.)	Tally	Frequency
1.0–1.9	|||	3
2.0–2.9	||||	4
3.0–3.9	~~||||~~	5
4.0–4.9	||||	4
5.0–5.9	||||	4
6.0–6.9	||	2
7.0–7.9	||	2

2. **BOOKS** The list below shows book prices for various books at a used book sale. Organize the data in a table using intervals $1.00–$1.99, $2.00–$2.99, $3.00–$3.99, and so on. What is the most common interval of book prices?

$3.78	$1.05	$6.52	$1.65	$4.99	$2.83
$1.52	$4.85	$4.64	$5.10	$3.09	$1.90
$6.29	$3.72	$6.50	$3.39	$2.55	$1.89
$3.22	$4.26	$5.29	$4.99	$1.10	$2.50

Use any strategy to solve Exercises 3–5. Some strategies are shown below.

PROBLEM-SOLVING STRATEGIES
• Use logical reasoning. • Act it out. • Make a table.

3. **ART FAIR** At the art fair, 95 artists exhibited their work. Of those 95 artists, 25 showed sculptures and 48 showed paintings. If 12 showed both sculptures and paintings, how many artists showed only sculptures or paintings?

4. **NUMBER CUBE** Jacy tossed a number cube several times and recorded the number shown after each toss. His results are listed below. Find the number that was tossed most frequently.

3	6	1	3	5	3	4	2	6
1	5	4	4	5	6	6	1	4
4	2	5	6	1	1	2	3	6

5. **GEOGRAPHY** Finland has a land area of 117,943 square miles. If the total area of Finland is 130,128 square miles, what percent of Finland's total area is water, to the nearest tenth of a percent?

NAME ______________________________ DATE ____________ PERIOD _____

11-2 Study Guide and Intervention

Histograms

Data from a frequency table can be displayed as a histogram. A **histogram** is a type of bar graph used to display numerical data that have been organized into equal intervals. To make a histogram from a frequency table, use the following steps.

Step 1 Draw and label a horizontal and a vertical axis. Include a title.

Step 2 Show the intervals from the frequency table on the horizontal axis.

Step 3 For each interval on the horizontal axis, draw a bar whose height is given by the frequencies.

Example **FOOTBALL The frequency table at the right shows the scores of all NFL teams in the first game of a recent season. Draw a histogram to represent the data.**

NFL Team Scores		
Score	Tally	Frequency
0–9	𝍸 II	7
10–19	𝍸 III	8
20–29	𝍸 𝍸 III	13
30–39	III	3
40–49	I	1

The histogram was created using the steps listed above. The horizontal axis is labeled "Score," the vertical axis is labeled "Number of Teams," and the histogram is titled "NFL Team Scores." The intervals are shown on the horizontal axis, and the frequencies are shown on the vertical axis. A bar is drawn in each interval to show the frequencies.

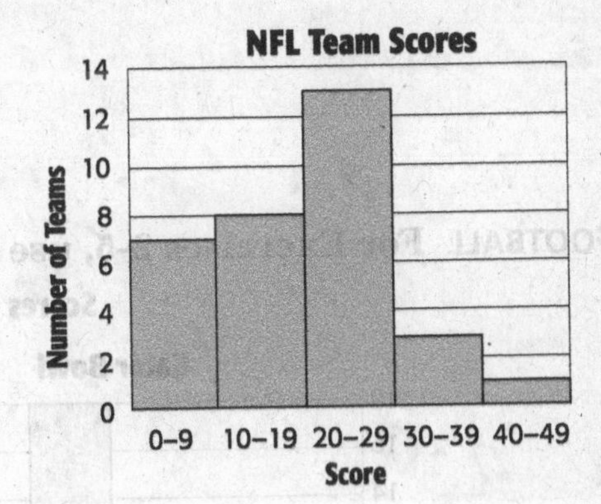

Exercise

TAXES The frequency table shows the tax on gasoline for the 50 states. Draw a histogram to represent the set of data.

Gas Tax for Each State		
Tax (cents/gal)	Tally	Frequency
8.1–12	II	2
12.1–16	𝍸	5
16.1–20	𝍸 𝍸 𝍸 𝍸 II	22
20.1–24	𝍸 𝍸 II	12
24.1–28	𝍸 I	6
28.1–32	III	3

NAME ______________________ DATE __________ PERIOD _____

11-2 Practice

Histograms

1. GOVERNMENT The list gives the year of birth for each state governor in the United States in 2007. Choose intervals and make a frequency table. Then construct a histogram to represent the data.

1944	1956	1952	1970	1957	1950
1964	1946	1942	1955	1941	1960
1957	1953	1955	1948	1963	1951
1946	1942	1963	1944	1940	1958
1947	1956	1956	1957	1944	1947
1956	1949	1954	1947	1942	1947
1946	1966	1960	1959	1954	1950
1935	1948	1947	1950	1943	

FOOTBALL **For Exercises 2–5, use the histograms shown.**

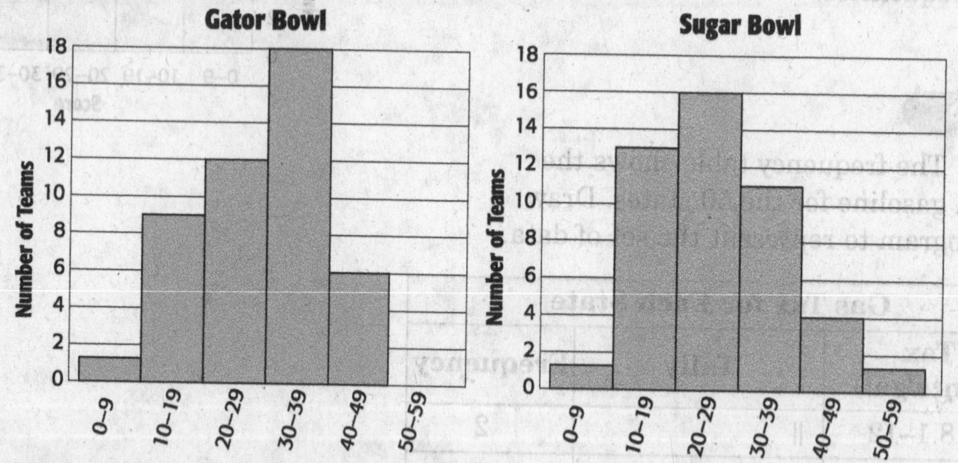

2. Which bowl game had the higher winning team score?
3. In which bowl game was the winning team score in the interval 30-39 points more often?
4. Determine which bowl game has had a winning team score of at least 30 points more often.
5. What was the lowest winning team score in each bowl game? Explain.

11-3 Study Guide and Intervention

Circle Graphs

A **circle graph** can be used to represent data that compares parts of a set of data to the whole set.

Example BASEBALL **Make a circle graph using the information in the table at the right.**

Batting in Major League Baseball	
Handedness	**Percent of Batters**
Right-Handed	62%
Left-Handed	26%
Switch Hitters /Unknown	12%

Step 1 There are 360° in a circle. So, multiply each percent by 360 to find the number of degrees for each section of the graph. Use a calculator.
Right-Handed: 62% of 360 = $0.62 \cdot 360$ or about 223°
Left-Handed: 26% of 360 = $0.26 \cdot 360$ or about 94°
Switch Hitters and Unknown: 12% of 360 = $0.12 \cdot 360$ or about 43°

Step 2 Use a compass to draw a circle and a radius. Then use a protractor to draw a 223° angle. This section represents right-handed hitters.

Step 3 From the new radius, draw a 94° angle. This section represents left-handed hitters. The remaining section represents switch hitters and players whose stance remains unknown. Now label each section. Then give the graph a title.

Batting in Major League Baseball

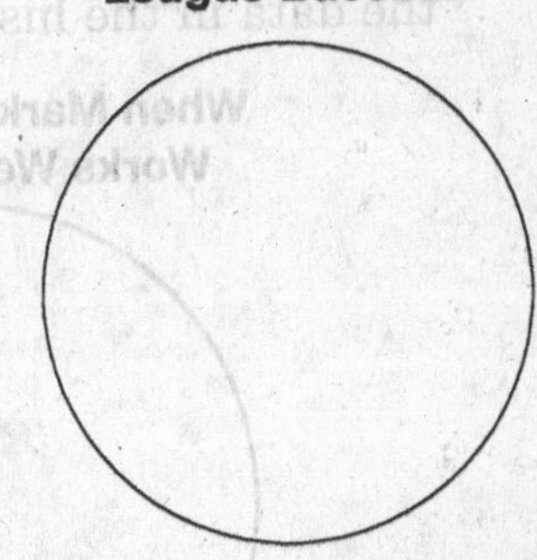

When the percents are not given, you must first determine what part of the whole each item represents.

Exercises

Make a circle graph for each set of data.

1.

U.S. Car and Truck Sales by Vehicle Size and Type	
Size and Type	**Percent**
Car	43%
Light Truck	54%
Heavy truck	3%

U.S. Car and Truck Sales by Vehicle Size and Type

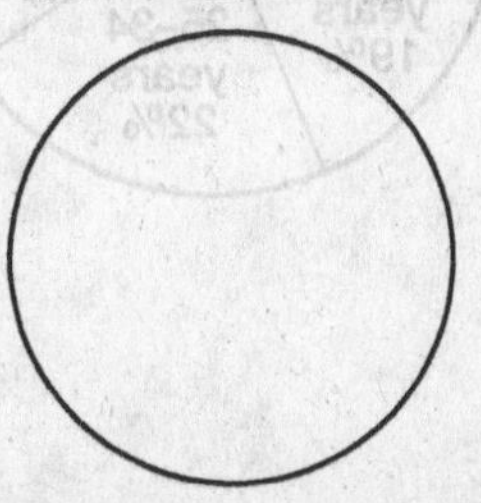

2.

Medals Won by the U.S. in the 2004 Summer Olympic Games	
Type	**Number**
Gold	35
Silver	39
Bronze	29

Medals Won by the U.S. in the 2004 Summer Olympic Games

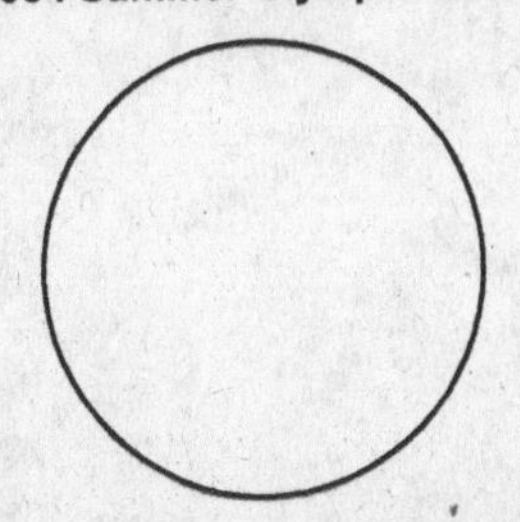

11-3 Practice

Circle Graphs

1. **ENERGY** Construct a circle graph for the data in the table.

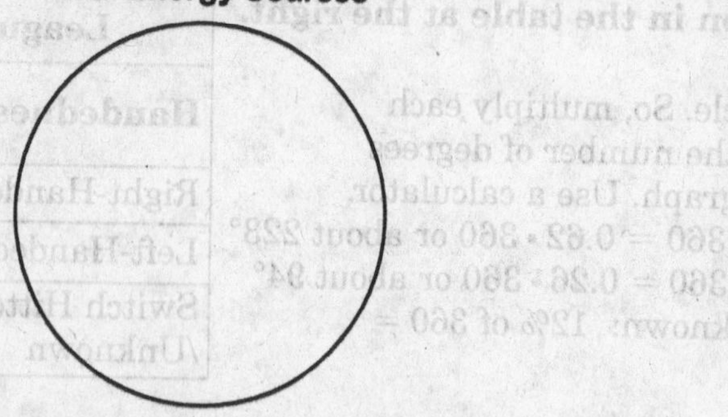

U.S. Energy Sources	
Petroleum	39.2%
Natural Gas	23.7%
Coal	22.8%
Nuclear Power	8.4%
Hydropower	2.7%
Other	3.2%

Source: *The World Almanac for Kids 2005*

2. **LITERATURE** Construct a circle graph for the data in the histogram.

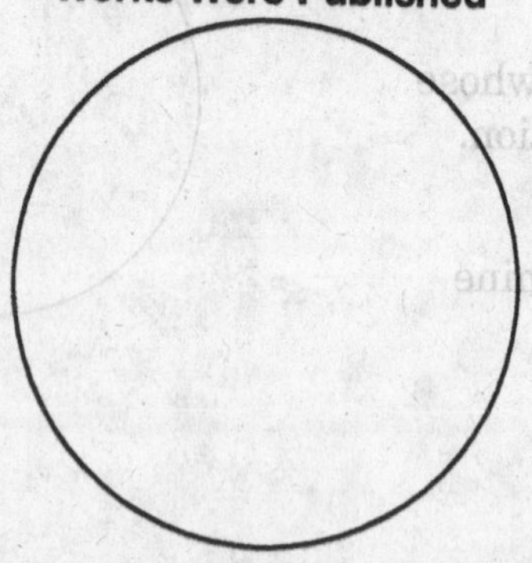

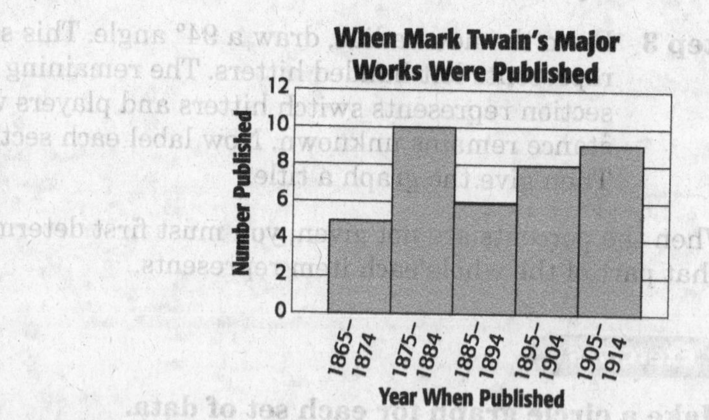

GOLF For Exercises 3 and 4, use the circle graph at the right.

3. Find the percent of golf equipment buyers who are 65 years or older. Then find the measure in degrees of the angle of the 65 years and older section of the graph to the nearest tenth degree.

4. Describe the buyers of golf equipment.

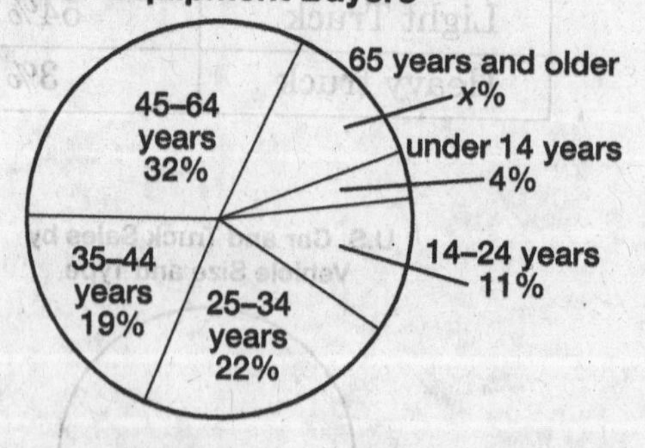

11-4 Study Guide and Intervention

Measures of Central Tendency and Range

The most common measures of central tendency are mean, median, and mode. The range is also used to describe a set of data. To find the **mean** of a data set, find the sum of the data values then divide by the number of items in the set. To find the **median** of a data set, put the values in order from least to greatest, then find the middle number. If there are two middle numbers, add them together and divide by 2. The **mode** of a data set is the number or numbers that occur most often. If no number occurs more than once, the data set has no mode. The **range** of a data set is the difference between the greatest number and the least number in a set of data.

Example **Find the mean, median, mode, and range of the set of data. Round to the nearest tenth if necessary. The ages, in years, of relatives staying at your home are listed below.**
5, 14, 8, 2, 89, 14, 10, 2

Mean $\frac{5 + 14 + 8 + 2 + 89 + 14 + 10 + 2}{8} = 18$

The mean age is 18.

Median Arrange the numbers in order from least to greatest.
2 2 5 8 10 14 14 89

The middle numbers are 8 and 10. Since $\frac{8 + 10}{2} = 9$, the median age is 9.

Mode The numbers 2 and 14 each occur twice. The data set has two modes, 2 and 14.

Range 89–2 or 87

Different circumstances determine which measure of central tendency or range is most appropriate to describe a set of data. The mean is most useful when the data has no extreme values. The median is most useful when the data has a few extreme values with no big gaps in the middle of the data. The mode is most useful when the data has many identical numbers.

Exercises

Find the mean, median, mode, and range of each set of data. Round to the nearest tenth if necessary.

1. 2, 4, 5, 1, 3

2. 7, 5, 7, 7, 6, 4

3. 18, 14, 15, 11, 14, 12, 17

4. 19, 24, 22, 16, 15, 27, 22, 27

5. 2.3, 1.1, 1.5, 3.2, 1.7, 2.0, 2.4, 1.8

6. 36, 32, 34, 34, 35, 38, 36, 34

7. 30, 29, 30, 31, 30

8. 4.2, 5.2, 2.3, 4.0, 4.6, 6.0, 2.3, 5.3

NAME ______________________ DATE ____________ PERIOD ______

11-4 Practice

Measures of Central Tendency and Range

Find the mean, median, mode, and range of each set of data. Round to the nearest tenth if necessary.

1. The prices, in dollars, of day packs
   37, 43, 41, 36, 43

2. Points on quizzes
   13, 6, 9, 8, 14, 5, 10, 7

3.

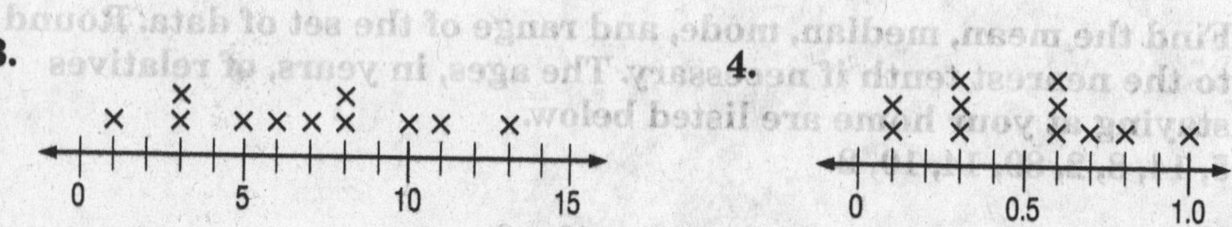

4.

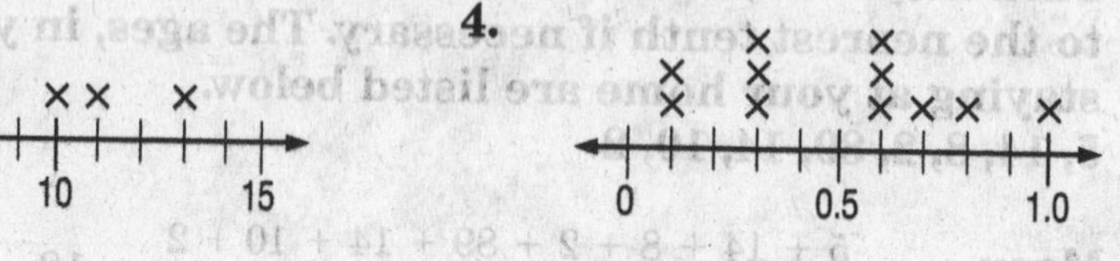

For Exercises 5 and 6, select the appropriate measure of central tendency or range to describe the data in each table. Justify your reasoning. Sample answers are given.

5.

Known Mountains on Mars	
Mountain	**Height (km)**
Alba Patera	3
Arsia Mons	9
Ascraeus Mons	11
Olympus Mons	27
Pavonis Mons	7

6.

Average Lengths of Wild Cats			
Cat	**Length**	**Cat**	**Length**
Cheetah	50.5 in.	Lion	102 in.
Eurasian Wildcat	24.3 in.	Puma	60 in.
Jaguar	57.5 in.	Serval	33.5 in.
Leopard	57 in.	Tiger	128 in.

Source: *Facts on File: Animal Fact File*

7. **MARS** Refer to the table of mountains on Mars in Exercise 5. Describe how the mean, median, mode, and range are each affected if the data for Olympus Mons is not included.

11-5 Study Guide and Intervention

Measures of Variation

The **lower quartile** or LQ is the median of the lower half of a set of data. The **upper quartile** or UQ is the median of the upper half of a set of data. The **interquartile range** is the difference between the upper quartile and the lower quartile.

Example 1 **Find the range, median, upper and lower quartiles, and interquartile range for the following set of data.**
13, 20, 18, 12, 21, 2, 18, 17, 15, 10, 14

The greatest number in the data set is 21. The least number is 2. The range is 21 − 2 or 19.

To find the quartiles, arrange the numbers in order from least to greatest.
2 10 12 13 14 15 17 18 18 20 21

The median is 15. The numbers below 15 are 2, 10, 12, 13, and 14. The median of the numbers below 15 is 12, so the lower quartile is 12. The numbers above 15 are 17, 18, 18, 20, and 21. The median of the numbers above 15 is 18, so the upper quartile is 18. The interquartile range is 18 − 12 or 6.

In some data sets, a few of the values are much greater than or less than the rest of the data. Data that are more than 1.5 times the value of the interquartile range beyond the quartiles are called **outliers**.

Example 2 **Find any outliers for the set of data given in Example 1.**

The interquartile range is 18−12 or 6.
Multiply the interquartile range by 1.5. $6 \times 1.5 = 9$

Any data more than 9 above the upper quartile or below the lower quartile are outliers. Find the limits of the outliers.

Subtract 9 from the lower quartile. $12 - 9 = 3$
Add 9 to the upper quartile. $18 + 9 = 27$

The limits of the outliers are 3 and 27. The only data point outside this range is 2, so the only outlier is 2.

Exercises

Find the range, median, upper and lower quartiles, interquartile range, and any outliers for each set of data.

1. 14, 16, 18, 24, 19, 15, 13

2. 29, 27, 24, 28, 30, 51, 28

3. 57, 60, 43, 55, 46, 43, 62, 31

4. 91, 92, 88, 89, 93, 95, 65, 85, 91

5. 104, 116, 111, 108, 113, 127, 109, 122, 115, 105

NAME ______________________________ DATE ____________ PERIOD ______

11-5 Practice

Measures of Variation

WILD CATS For Exercises 1–4, use the data in the table.

Average Birth Weights of Wild Cats			
Cat	**Weight (oz)**	**Cat**	**Weight (oz)**
Cheetah	7.5	Lion	48
Eurasian Wildcat	1.4	Puma	12
Jaguar	28	Serval	8.5
Leopard	17.5	Tiger	40

Source: *Facts on File: Animal Fact File*

1. What is the range of the data?

2. Find the median, the upper and lower quartiles, and the interquartile range of the data.

3. Identify any outliers.

4. Use the measures of variation to describe the data in the table.

WORD For Exercises 5–8, use the data in the table.

5. What is the range of the data?

6. Find the median, the upper and lower quartiles, and the interquartile range of the data.

Death Valley Average Monthly Precipitations			
0.19	0.13	0.35	0.12
0.12	0.05	0.42	0.18
0.11	0.42	0.14	0.10

Source: weather.com

7. Identify any outliers.

8. Use the measures of variation to describe the data in the stem-and-leaf plot.

11-6 Study Guide and Intervention

Box-and-Whisker Plots

A **box-and-whisker plot** uses a number line to show the distribution of a set of data. The *box* is drawn around the quartile values, and the *whiskers* extend from each quartile to the extreme data points that are not outliers.

Example **Use the data below to construct a box-and-whisker plot.**
12, 14, 8, 10, 1, 16, 10, 11, 10

Step 1 Put the data in order from least to greatest and find the median, lower quartile, upper quartile, and the least and greatest values that are not outliers.
Ordered data: 1, 8, 10, 10, 10, 11, 12, 14, 16

Least value: 1; Median: 10; Greatest value: 16;

Lower quartile: $\frac{8 + 10}{2}$ or 9; Upper quartile: $\frac{12 + 14}{2}$ or 13;

Interquartile range: 13 − 9 or 4;
Lower limit for outliers: 9 − 6 or 3;
Upper limit for outliers: 13 + 6 or 19;
Outliers: 1

Step 2 Draw a number line that includes the least and greatest numbers in the data.

Step 3 Mark the extremes, the median, and the upper and lower quartile above the number line. Since the data have an outlier, mark the least value that is not an outlier.

Step 4 Draw the box and the whiskers.

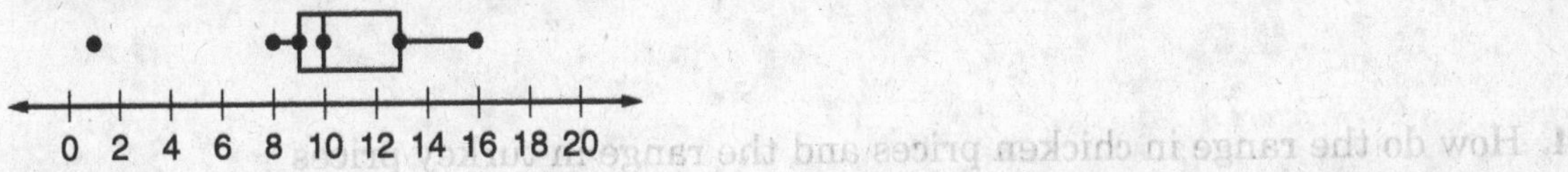

Box-and-whisker plots separate data into four parts. Even though the parts may differ in length, each part contains $\frac{1}{4}$ of the data.

Exercises

Construct a box-and-whisker plot for each set of data.

1. 4, 7, 5, 3, 9, 6, 4

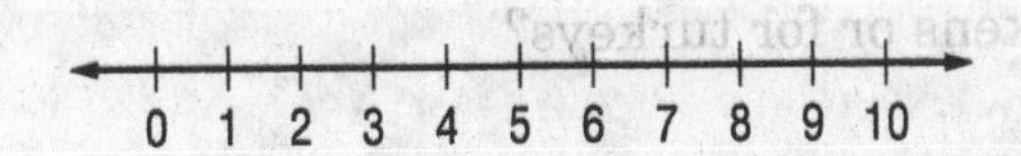

2. 13, 12, 17, 10, 6, 11, 14

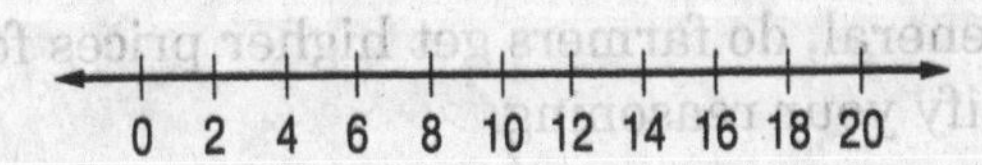

3. 23, 36, 22, 34, 30, 29, 26, 27, 33

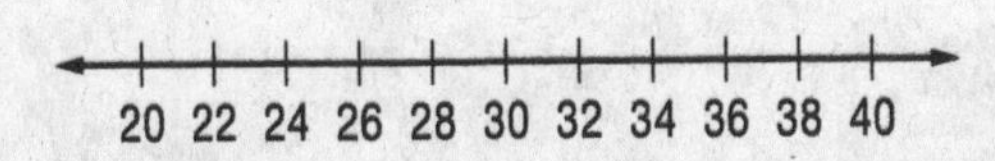

4. 108, 126, 110, 104, 106, 123, 140, 122, 114, 109

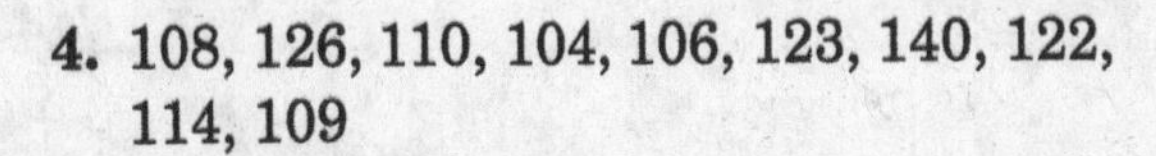
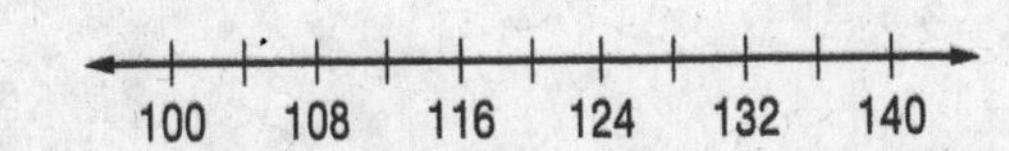

NAME ______________________________ DATE ____________ PERIOD _____

11-6 Practice

Box-and-Whisker Plots

Construct a box-and-whisker plot for each set of data.

1. Ages of children already signed up for swimming classes: 10, 12, 9, 7, 10, 12, 14, 14, 10, 16

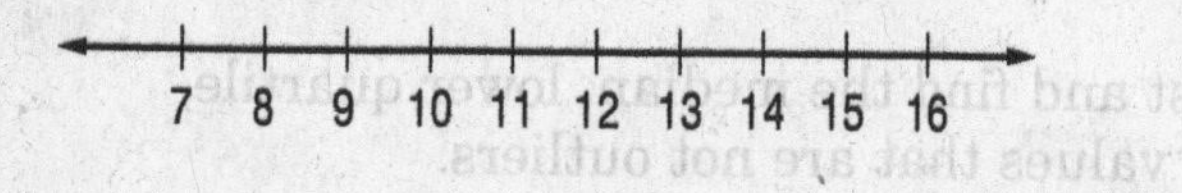

2. Prices, in dollars, of bicycles on sale: 150, 134, 132, 120, 145, 170, 125, 130, 145, 185, 140

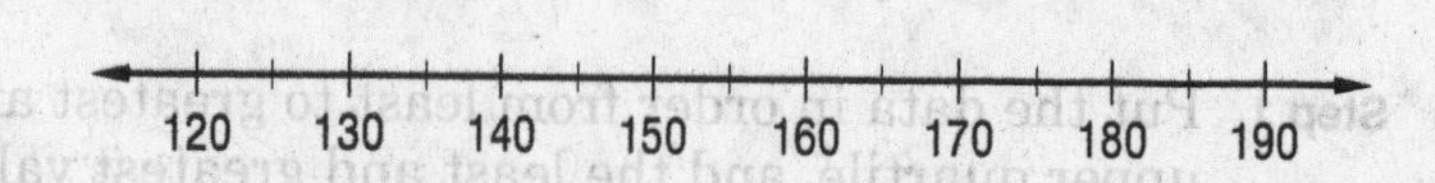

POULTRY For Exercises 3–7, use the box-and-whisker plot below.

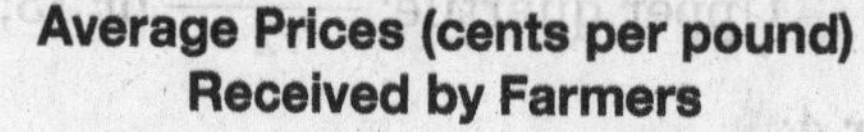

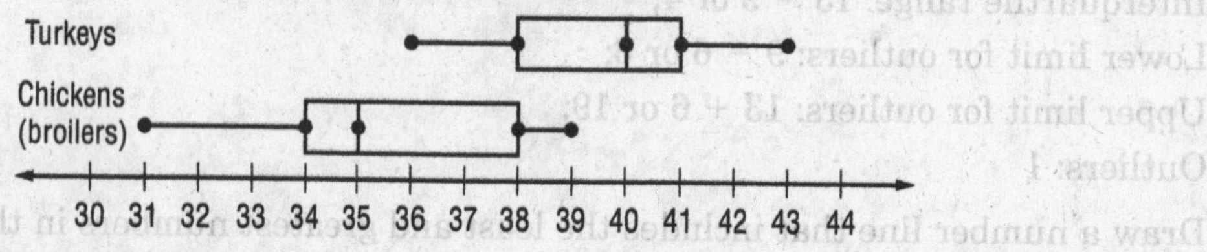

Source: *The World Almanac*

3. How many outliers are in the data?

4. How do the range in chicken prices and the range in turkey prices compare?

5. In the chicken prices, which quartiles show the greatest spread of data?

6. What percent of the data for the turkey prices is above the upper quartile for the chicken prices?

7. In general, do farmers get higher prices for chickens or for turkeys? Justify your reasoning.

NAME ______________________________ DATE ____________ PERIOD _____

11-7 Study Guide and Intervention

Stem-and-Leaf Plots

Stem-and-Leaf Plot	
	Words One way to organize and display data is to use a **stem-and-leaf plot**. In a stem-and-leaf plot, numerical data are listed in ascending or descending order.
	Model (see below)

Stem	Leaf
2	0 1 1 2 3 5 5 6
3	1 2 2 3 7 9
4	0 3 4 8 8

3 | 7 = 37

The greatest place value of the data is used for the **stems**.

The next greatest place value forms the **leaves**.

Example **ZOOS Display the data shown at the right in a stem-and-leaf plot.**

Size of U. S. Zoos	
Zoo	Size (acres)
Audubon (New Orleans)	58
Cincinnati	85
Dallas	95
Denver	80
Houston	55
Los Angeles	80
Oregon	64
St. Louis	90
San Francisco	75
Woodland Park (Seattle)	92

Step 1 The least and the greatest numbers are 55 and 95. The greatest place value digit in each number is in the tens. Draw a vertical line and write the stems from 5 to 9 to the left of the line.

Stem	Leaf
5	8 5
6	4
7	5
8	5 0 0
9	5 0 2

Step 2 Write the leaves to the right of the line, with the corresponding stem. For example, for 85, write 5 to the right of 8.

Step 3 Rearrange the leaves so they are ordered from least to greatest. Then include a key or an explanation.

Stem	Leaf
5	5 8
6	4
7	5
8	0 0 5
9	0 2 5

8 | 5 = 85 acres

Exercises

Display each set of data in a stem-and-leaf plot.

1. {27, 35, 39, 27, 24, 33, 18, 19}

2. {94, 83, 88, 77, 95, 99, 88, 87}

ROLLER COASTERS For Exercises 3 and 4, use the stem-and-leaf plot shown.

The Fastest Roller Coasters

Stem	Leaf
8	3 5
9	2 5
10	0

8 | 3 = 83 mph

3. What is the speed of the fastest roller coaster? The slowest?

4. What is the median speed?

NAME ______________________________ DATE ____________ PERIOD _____

11-7 Practice

Stem-and-Leaf Plots

Display each set of data in a stem-and-leaf plot.

1. {68, 63, 70, 59, 78, 64, 68, 73, 61, 66, 70}

2. {27, 32, 42, 31, 36, 37, 47, 23, 39, 31, 41, 38, 30, 34, 29, 42, 37}

3.

Major League Baseball Leading Pitchers, 2007	
Player	**Wins**
J. Beckett	20
F. Carmona	19
J. Lackey	19
B. Webb	18
A. Harang	16
T. Hudson	16
K. Escobar	18
T. Wakefield	17
J. Peavy	19
J. Francis	17

4.

Average Prices Received by U.S. Farmers	
Commodity	**Price (dollars per 100 pounds)**
Beef Cattle	86
Hogs	49
Lambs	101
Milk	16
Veal Calves	119

Source: *U. S. Department of Agriculture*

RECREATION For Exercises 5–7, use the information in the back-to-back stem-and-leaf plot shown at the right.

5. The category with the lowest total expenditure in 2002 was motion pictures. What was its total?

6. What is the median total recreational spending for 2002? For 2006?

7. Compare the total spending on recreation in 2002 with that in 2006.

Total U.S. Spending on Personal Recreation (by Category)

2002		2006
7 5 5	0	9
8 7 2 0	1	0 2 8
7 2	2	2
4 0	3	4 5 7
	4	4
	5	6
1	6	0
	7	
	8	4
	9	

7 | 2 = $27 billion *3 | 5 = $35 billion*

11-8 Study Guide and Intervention

Select an Appropriate Display

There are many different ways to display data. Some of these displays and their uses are listed below.

Type of Display	Best Used to
Bar Graph	show the number of items in specific categories.
Box-and-Whisker Plot	show measures of variation for a set of data.
Circle Graph	compare parts of the data to the whole.
Histogram	show frequency of data divided into equal intervals.
Line Graph	show change over a period of time.
Line Plot	show how many times each number occurs in the data.
Stem-and-Leaf Plot	list all individual numerical data in condensed form.
Venn Diagram	show how elements among sets of data are related.

As you decide what type of display to use, ask the following questions.

- What type of information is this?
- What do I want my graph or display to show?

Remember, all data sets can be displayed in more than one way. And there is often more than one appropriate way to display a given set of data.

Examples **Choose an appropriate type of display for each situation.**

1 the change in the winning times for the Kentucky Derby for the last 15 years

This data does not deal with categories or intervals. It deals with the change of a value over time. A line graph is a good way to show changes over time.

2 energy usage in the U.S., categorized by the type of user

In this case, there are specific categories. If you want to show the specific amount of energy used in each category, use a bar graph. If you want to show how each category is related to the whole, use a circle graph.

Exercises

Select an appropriate type of display for each situation. Justify your reasoning.

1. the cost of homeowners insurance over the past 10 years

2. the amount of federally owned land in each state, arranged in intervals

NAME ______________________________ DATE ____________ PERIOD _____

11-8 Practice

Select an Appropriate Display

Select an appropriate type of display for each situation. Justify your reasoning.

1. prices of athletic shoes in the store arranged by intervals

2. the numbers of teens who spend Saturdays doing homework, playing, and/or doing chores

3. the number of each of four kinds of trees found in the forest

4. the spread of the run times for the first $\frac{1}{4}$ of the runners completing a marathon

Select an appropriate type of display for each situation. Justify your reasoning. Then construct the display.

5.

Heights of Mountains on the Moon	
Height	**Percent of the Mts.**
Less than 1 km	11.8%
1-2 km	17.7%
2-3 km	17.7%
3-4 km	35.3%
More than 4 km	17.7%

6. **WORK** Jim worked 1 hour on Monday. On Tuesday, he worked 2 more hours than he worked on Monday. On Wednesday, he worked 2 more hours than he worked on Tuesday. The pattern continued through Friday.

NAME ______________________ DATE ____________ PERIOD ______

12-1

Study Guide and Intervention

Counting Outcomes

An organized list of outcomes, called a **sample space**, can help you determine the total number of possible outcomes for an event.

Example 1 COMPUTERS **An electronics store offers a model 2010 processor with a choice of 2 monitors (15-inch and 17-inch) and 2 printers (inkjet and laser). Draw a tree diagram to determine how many different computer systems are available.**

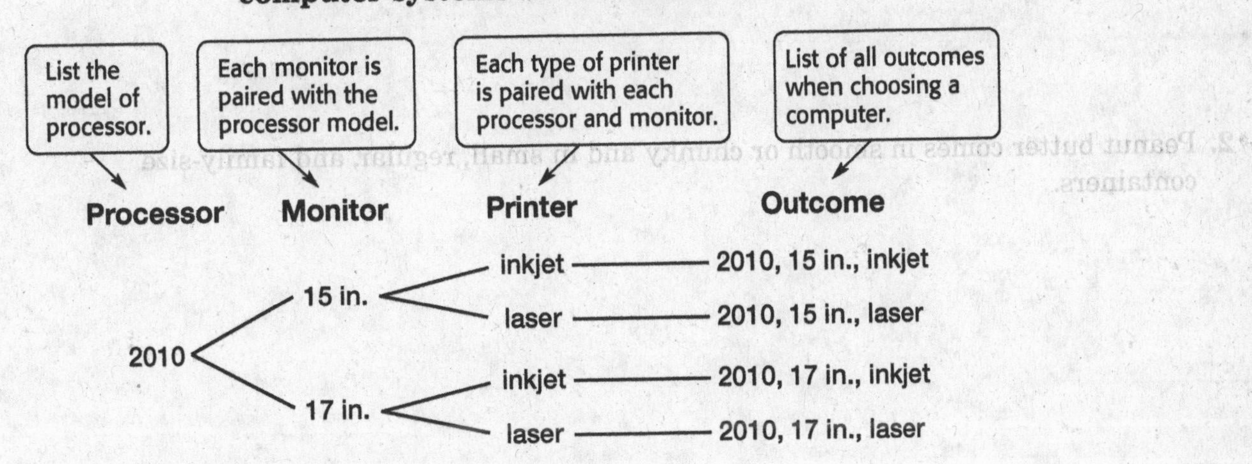

There are 4 different computer systems available.

If event *M* can occur in *m* ways and is followed by event *N* that can occur in *n* ways, then the event *M* followed by the event *N* can occur in $m \cdot n$ ways. This principle is known as the **Fundamental Counting Principle.**

Example 2 LOCKS **A lock combination is made up of three numbers from 0 to 39. How many combinations are possible?**

Use the Fundamental Counting Principle. $40 \times 40 \times 40 = 64{,}000$

There are 64,000 possible lock combinations.

Exercises

1. A museum tour includes a box lunch which contains a ham, turkey, or cheese sandwich and an apple, a banana, an orange, or a pear. An equal number of all lunch combinations are available for each tour. Draw a tree diagram to determine the number of outcomes.

Use the Fundamental Counting Principle to find the number of possible outcomes.

2. A number cube is rolled twice.

3. Six coins are tossed.

NAME ______________________ DATE ____________ PERIOD ______

Practice

Counting Outcomes

Draw a tree diagram to determine the number of possible outcomes.

1. A coin is tossed and one of the vowels A, E, I, O, and U is chosen at random.

2. Peanut butter comes in smooth or chunky and in small, regular, and family-size containers.

Use the Fundamental Counting Principle to find the number of possible outcomes.

3. A month of the year is picked at random and a coin is tossed.
4. A quarter and a dime are tossed and a number cube is rolled.
5. There are 8 true-false questions on a science quiz.

GIFT BASKETS For Exercises 6–9, use the following information.

Gina made gift baskets to sell at her electronics store. Each basket had a CD of either classical or jazz music, a DVD of a comedy or an action film, a video game or a financial software package, and a small, medium, or large T-shirt with her store's logo. An equal number of baskets of each possible combination were made.

6. How many different basket combinations were there?
7. Find the number of combinations that had a video game.
8. If a basket was chosen randomly, what is the probability that the basket contained an action film?
9. What is the probability of choosing a basket with classical music and a large T-shirt?

NAME ______________________ DATE ____________ PERIOD ______

Study Guide and Intervention

Probability of Compound Events

The probability of two independent events can be found by multiplying the probability of the first event by the probability of the second event.

Example 1 **Two number cubes, one red and one blue, are rolled. What is the probability that the outcome of the red number cube is even and the outcome of the blue number cube is a 5?**

$P(\text{red number cube is even}) = \frac{1}{2}$

$P(\text{blue number cube is a 5}) = \frac{1}{6}$

$P(\text{red number cube is even and blue number cube is a 5}) = \frac{1}{2} \cdot \frac{1}{6}$ or $\frac{1}{12}$

The probability that the two events will occur is $\frac{1}{12}$.

If two events, *A* and *B*, are dependent, then the probability of both events occurring is the product of the probability of *A* and the probability of *B* after *A* occurs.

Example 2 **There are 6 black socks and 4 white socks in a drawer. If one sock is taken out without looking and then a second is taken out, what is the probability that they both will be black?**

$P(\text{first sock is black}) = \frac{6}{10}$ or $\frac{3}{5}$ — 6 is the number of black socks; 10 is the total number of socks.

$P(\text{second sock is black}) = \frac{5}{9}$ — 5 is the number of black socks after one black sock is removed; 9 is the total number of socks after one black sock is removed.

$P(\text{two black socks}) = \frac{3}{5} \cdot \frac{5}{9}$ or $\frac{1}{3}$

The probability of choosing two black socks is $\frac{1}{3}$.

Exercises

A card is drawn from a deck of 10 cards numbered 1 through 10 and a number cube is rolled. Find each probability.

1. *P*(10 and 3)

2. *P*(two even numbers)

3. *P*(two prime numbers)

4. *P*(9 and an odd number)

5. *P*(two numbers less than 4)

6. *P*(two numbers greater than 5)

There are 4 red, 6 green, and 5 yellow pencils in a jar. Once a pencil is selected, it is not replaced. Find each probability.

7. *P*(red and then yellow)

8. *P*(two green)

9. *P*(green and then yellow)

10. *P*(red and then green)

NAME ______________________ DATE ____________ PERIOD ________

Practice

Probability of Compound Events

The two spinners at the right are spun. Find each probability.

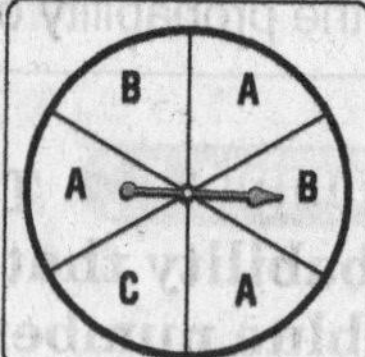

1. P(4 and C)
2. P(1 and A)
3. P(even and C)
4. P(odd and A)
5. P(greater than 3 and B)
6. P(less than 5 and B)

GAMES **There are 10 yellow, 6 green, 9 orange, and 5 red cards in a stack of cards turned facedown. Once a card is selected, it is not replaced. Find each probability.**

7. P(two yellow cards)
8. P(two green cards)
9. P(a yellow card and then a green card)
10. P(a red card and then an orange card)
11. P(two cards that are not orange)
12. P(two cards that are neither red nor green)
13. OFFICE SUPPLIES A store sells a box of highlighters that contains 4 yellow, 3 blue, 2 pink, and 1 green highlighter. What is the probability of randomly picking 1 blue and 1 pink highlighter from the box?
14. BASKETBALL Angelina makes 70% of her free throws. What is the probability that she will make her next two free throws?

CAR RENTALS **For Exercises 15 and 16, use the following information and the information in the table.**

At a car rental office, 63% of the customers are men and 37% are women.

Car Requests	
Compact	25%
Full-size	37%
Convertible	10%
SUV	16%
Luxury	12%

15. What is the probability that the next customer will be a woman who requests a convertible?
16. What is the probability that the next customer will be a man who requests either a compact car or luxury car?

NAME ________________________ DATE ____________ PERIOD ______

12-3 Study Guide and Intervention

Experimental and Theoretical Probability

Probabilities based on the outcomes obtained by conducting an experiment are called **experimental probabilities.** Probabilities based on known characteristics or facts are called **theoretical probabilities.** Theoretical probability tells you what *should* happen in an experiment.

Examples **Kuan is conducting an experiment to find the probability of getting 0, 1, 2, or 3 heads when tossing three coins on the floor. The results of his experiment are given at the right.**

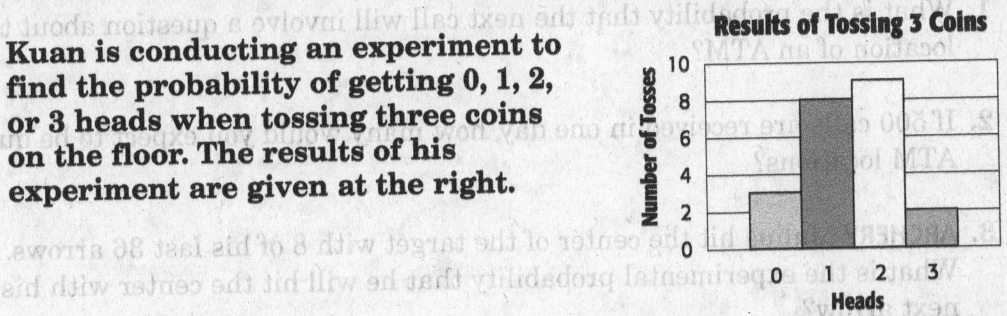

1 **Based on the results in the bar graph, what is the probability of getting 3 heads on the next toss?**

There were 22 tosses and 2 of those had 3 heads. The experimental probability is $\frac{2}{22}$ or $\frac{1}{11}$.

2 **Based on the experimental probability, how many times should Kuan expect to get 3 heads in the next 55 tosses?**

Kuan should expect to get 3 heads about $\frac{1}{11} \cdot 55$ or 5 times.

3 **What is the theoretical probability of getting 3 heads on a toss?**

The theoretical probability is $\frac{1}{2} \cdot \frac{1}{2} \cdot \frac{1}{2}$ or $\frac{1}{8}$.

The experimental probability and the theoretical probability seem to be consistent.

Exercises

Use the table that shows the results of spinning a game spinner 50 times.

Color	Number of Times
green	18
red	24
blue	8

1. Based on the results in the table, what is the probability of spinning green?

2. Based on the results, how many green spins would you expect to occur in 300 spins?

3. What is the theoretical probability of spinning green?

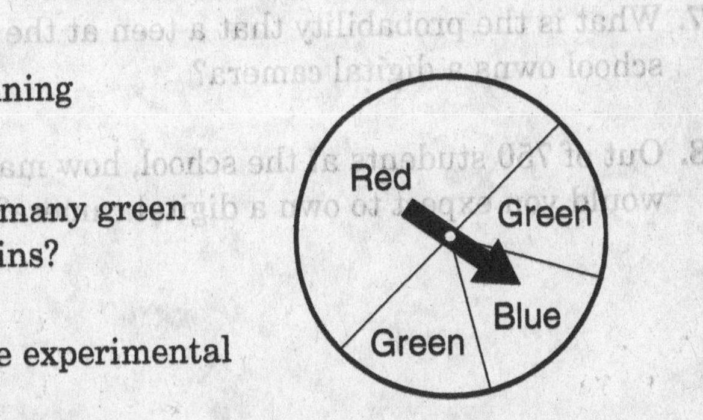

4. Based on the theoretical probability, how many green spins would you expect to occur in 300 spins?

5. Compare the theoretical probability to the experimental probability.

12-3

NAME ____________________ DATE __________ PERIOD _____

Practice

Experimental and Theoretical Probability

TELEPHONES For Exercises 1 and 2, use the following information.
Of the last 45 telephone calls received at a bank, 6 involved questions about Automatic Teller Machines (ATM) locations.

1. What is the probability that the next call will involve a question about the location of an ATM?

2. If 500 calls are received in one day, how many would you expect to be questions about ATM locations?

3. **ARCHERY** Julius hit the center of the target with 8 of his last 36 arrows. What is the experimental probability that he will hit the center with his next arrow?

DRINKS For Exercises 4 and 5, use the information about drinks ordered by 200 customers at a restaurant.

4. What is the probability that a customer ordered milk?

5. On a day when the restaurant has 800 customers, how many would you expect to order milk?

Drinks Ordered	
Drink	**Number**
Water	64
Milk	22
Coffee	35
Soft Drink	68
Other	11

6. **NEWSPAPERS** In the last 40 days, Mr. Neptune's newspaper has been delivered late 6 times. What is the experimental probability that it will be delivered late tomorrow?

TECHNOLOGY For Exercises 7 and 8, use the results of a survey of 80 teens at a school shown at the right.

7. What is the probability that a teen at the school owns a digital camera?

8. Out of 750 students at the school, how many would you expect to own a digital camera?

Technology Device	Number Who Own
Cell Phone	45
Digital Camera	32
DVD Player	65
Laptop Computer	18

NAME ______________________ DATE __________ PERIOD ______

12-4

Study Guide and Intervention

Problem-Solving Investigation: Act It Out

Example **Michael has a red square tile, a blue square tile, a green square tile, and a yellow square tile. How many different ways can he arrange the tiles so that they form a larger square?**

Understand There are four tiles that can be arranged into a larger 2 by 2 square. How many different ways can the tiles be arranged into the larger square?

Plan Use letters to stand for each color tile. Arrange the tiles starting with each combination of tiles that has the red tile in the upper left corner. Then repeat this step for each of the other three colors.

Solve

RB GY	RB YG	RG BY	RG YB	RY BG	RY GB	There are 6 large squares with the red tile in the upper left.
BR GY	BR YG	BG RY	BG YR	BY RG	BY GR	There are 6 large squares with the blue tile in the upper left.
GR BY	GR YB	GB RY	GB YR	GY RB	GY BR	There are 6 large squares with the green tile in the upper left.
YR BG	YR GB	YB RG	YB GR	YG RB	YG BR	There are 6 large squares with the yellow tile in the upper left.

Check Each larger square with the red square in the upper left corner is shown for a total of six. Therefore there should be 6 sets for each color. $4 \times 6 = 24$. There are 24 ways that Michael can arrange the tiles into larger squares.

Exercises

For Exercises 1–3, solve each problem using the act it out strategy.

1. **GEOMETRY** How many different pairs of regular polygons can be made from 16 toothpicks with none left over if only one toothpick is used for each side?

2. **MONEY** Byron wants to buy a comic book that costs $0.65. If he uses exact change, how many different combinations of nickels, dimes, and quarters can he use?

3. **NUMBER LINE** In a math class game, players are using a number line on the floor. Grace starts at zero and moves forward 7 numbers on her first turn and moves backward 4 numbers on her second turn. If this pattern continues, how many turns will it take for her to move forward to 16?

12-4

NAME ______________________ DATE ____________ PERIOD ______

Practice

Problem-Solving Investigation: Act It Out

Mixed Problem Solving

For Exercises 1 and 2, use the act it out strategy.

1. **BILLS** Joaquin bought a DVD for $21. He gave the cashier two $20 bills. How many different combinations of $1, $5, and $10 bills can the cashier give him for change?

2. **TENNIS** Felix, Lolita, Tetsuo, Ling, and Maxine are on the school tennis team. When ranked from first to fifth, how many ways can they be ranked if Maxine is always first and Felix is always ranked above Tetsuo?

Use any strategy to solve Exercises 3–6. Some strategies are shown below.

PROBLEM-SOLVING STRATEGIES
• Work backward. • Look for a pattern. • Use logical reasoning. • Act it out.

3. **PUMPKINS** Mr. Greene harvested pumpkins for selling at four markets. He sold one-fifth of his crop at the first market, 40 at the second, 25% of the remaining at the third, and twice what he sold at the second at the forth market. If Mr. Greene has one pumpkin remaining, how many pumpkins did he sell?

4. **CHORES** Kimberley has the choice of washing the car, mowing the lawn, or raking leaves on Saturday and baking a cake, washing the dishes, or doing the laundry on Sunday. In how many ways can she choose one chore for each day?

5. **FUNDRAISER** The drama club is selling 100 T-shirts for $15 each for a fundraiser. The T-shirts cost a total of $623. If they sell all the T-shirts, how much money will be raised for the drama club?

6. **DELICATESSEN** A delicatessen offers the possibility of 180 kinds of sandwiches made with wheat, rye, white, or sourdough breads. If the delicatessen adds multi-grain bread to the menu, find the number of possible kinds of sandwiches the delicatessen now offers.

12-5

NAME ______________________________ DATE ______________ PERIOD ______

Study Guide and Intervention

Using Sampling to Predict

Data gathered from a representative sample can be used to make predictions about a population. An **unbiased sample** is selected so that it is representative of the entire population. In a **biased sample,** one or more parts of the population are favored over others.

Examples **Describe each sample.**

1 **To determine the favorite dog breed of people who enter dog shows, every fifth person entering a dog show is surveyed.**

Since the people are selected according to a specific pattern, the sample is a systematic random sample. It is an unbiased sample.

2 **To determine what type of pet people prefer, the spectators at a dog show are surveyed.**

The spectators at a dog show probably prefer dogs. This is a biased sample. The sample is a convenience sample since all of the people surveyed are in one location.

Examples COOKIES **Students in the eighth grade surveyed 50 students at random about their favorite cookies. The results are in the table at the right.**

Flavor	Number
oatmeal	15
peanut butter	11
chocolate chip	16
sugar	8

3 **What percent of students prefer chocolate chip cookies?**

16 out of 50 students prefer chocolate chip cookies.

$16 \div 50 = 0.32$ 32% of the students prefer chocolate chip cookies.

4 **If the students order 500 boxes of cookie dough, how many boxes should be chocolate chip?**

Find 32% of 500.

$0.32 \times 500 = 160$ About 160 boxes of cookie dough should be chocolate chip.

Exercises

Describe the sample.

1. To determine if the tomatoes in 5 boxes stacked on a pallet are not spoiled, the restaurant manager checks 3 tomatoes from the top box.

A random survey of the students in eighth grade shows that 7 prefer hamburgers, 5 prefer chicken, and 3 prefer hot dogs.

2. What percent prefer hot dogs?

3. If 120 students will attend the eighth grade picnic, how many hot dogs should be ordered?

Lesson 12-5

Practice

Using Sampling to Predict

Determine if each conclusion is valid. Then describe each sample.

1. To determine the most common injury cared for in an emergency room, a reporter goes to the same hospital every afternoon for one month during the summer and observes people entering the emergency room. She concludes that second degree sunburn is the most common injury.

2. To evaluate customer satisfaction, a grocery store gives double coupons to anyone who completes a survey as they enter the store. The store manager determines that customers are very satisfied with their shopping experience in his store.

3. To evaluate the integrity of underground water lines, the department of public works randomly selects 20 sites in each of 5 sectors of the city to unearth and observe the water lines. At 24 of the sites, the water lines needed repair. The department of public works concludes that one-fourth of underground water lines throughout the city need repair.

4. **DOWNLOADS** A guidance counselor asked students who owned mobile phones, which was the last type of download each one downloaded to their mobile phone. The results are shown in the table. If there are 420 students in the school, how many can be expected to download *ringtones*?

Mobile Phone Downloads

Type	Frequency
Games	10
Ringtones	25
Screensavers	14
Music	36

DENTISTRY A survey is to be conducted to determine the reasons dental patients are hesitant to go to the dentist. Describe the sample and explain why each sampling method might not be valid.

5. Adults that are randomly selected from an office complex are asked to go online and fill out a questionnaire.

6. A randomly selected dentist asks his patients why they may be hesitant to go to a dentist.

7. Randomly selected dental patients from different age groups who are having a routine check-up are asked to write down their feelings.